D0622912

GOD IS WATCHING YOU

GOD IS WATCHING YOU

HOW THE FEAR OF GOD MAKES US HUMAN

DOMINIC JOHNSON

OXFORD
UNIVERSITY PRESS

OXFORD
UNIVERSITY PRESS

Oxford University Press is a department of the University of
Oxford. It furthers the University's objective of excellence in research,
scholarship, and education by publishing worldwide.
Oxford is a registered trade mark of Oxford University Press in the UK
and in certain other countries

Published in the United States of America by Oxford University Press
198 Madison Avenue, New York, NY 10016, United States of America

Library of Congress Cataloging-in-Publication Data
Johnson, Dominic (Professor of International Relations)
God is watching you : how the fear of God makes us human / Dominic Johnson.
pages cm
Includes bibliographical references and index.
ISBN 978–0–19–989563–2 (cloth : alk. paper)   1. Religion—Philosophy.   2. Fear of
God.   3. Punishment—Religious aspects.   I. Title.
BL51.J675 2015
202'.3—dc23
2015009949

1 3 5 7 9 8 6 4 2
Printed in the United States of America
on acid-free paper

CONTENTS

ACKNOWLEDGEMENTS

One afternoon in Uganda in 1994, Oliver Krueger jumped off the back of our truck to get a low level photo of a beautiful male lion that was gazing at us intently through the tall grass a short distance away. Seeing this unkempt German wielding a massive 500 mm telephoto lens on the end of his camera, the lion perfectly reasonably decided to charge, issuing a booming, bone-shaking roar as it bounded toward us with effortless leaps. The driver of the truck, unaware of Oliver's disembarkation at the back, hit the gas and the truck sped off in a cloud of dust, leaving our seemingly doomed friend sprinting along the track behind us. We banged madly on the roof, and the truck screeched to a halt. Oliver covered 100 m in what seemed like 3 seconds and clambered on board, camera in one piece, nerves shattered. The lion drew up, his point made. There are some things that human beings don't argue with, even if they're only making a threat.

I'm glad Oliver didn't get eaten that day, because among the many things I gained from our long friendship in Africa and later in Oxford was inspiration for our original paper laying out an evolutionary perspective on the fear of God and human cooperation, entitled "The Good of Wrath." Evolution, it seemed to us, had good reasons to make use of our fear.

I am grateful for the support of the Kennedy Memorial Trust and Harvard University, where that paper was completed, with valuable help and advice from Nick Brown, Mark Molesky, Stephen Peter Rosen, Monica Duffy Toft, Luis Zaballa, and Richard Wrangham. Special thanks are also due for the support of Branco Weiss and the Society in Science program at ETH Zurich, Switzerland, and to Leonard Barkan,

Mary Harper, Simon Levin, and the Society of Fellows at Princeton University where I spent three heavenly years writing a series of subsequent papers that developed the supernatural punishment idea further. Further thanks are due to my co-author Jesse Bering during those years, who brought remarkable insights from human cognition that powerfully bolstered the theory, and to the John Templeton Foundation for a generous grant on the adaptive logic of religious beliefs and behavior.

I am particularly indebted for a wonderful residential research year that allowed me to complete this book at the Center of Theological Inquiry back at Princeton. William Storrar and Robin Lovin had the vision to bring together a group of scientists, philosophers, and theologians to think seriously about the interactions between religion and science in understanding human origins, and were extremely generous in their support for this project and the writing of the manuscript. We were assisted in these efforts by unerring help from Shirah Metzigian, Carlee Beard, and Linda Arntzenius. I am especially grateful for the intellectual input from the assembled fellows that year, many of whom spoke a very different language, but ones that were vital for me to learn. Thanks to Lee Cronk, Conor Cunningham, Celia Deane-Drummond, Agustin Fuentes, Hillary Lenfesty, Jan-Olav Henriksen, Nicola Hoggard-Creegan, Markus Muehling, Eugene Rogers, Robert Song, and Aku Visala. I continue to hear your voices when I write—which is sometimes heaven and sometimes hell.

A special mention is reserved for Jeff Schloss and Richard Sosis, also two of the Princeton fellows and long standing colleagues in the field who have repeatedly surpassed my expectations of human nature. Over many years, they have displayed such an incredible degree of intellectual openness, goodwill, and generosity of spirit that sometimes I doubted the logic of natural selection—at least in their lineage.

I am indebted to Ara Norenzayan and Azim Shariff, who have painstakingly conducted so much brilliant empirical work on the cognitive science of religion and the role of supernatural surveillance and punishment on prosocial behavior. Without their work, it would still be largely just a theory. The Oxford University Press marketing department (not me) chose the same title for this book as their seminal paper on religious primes and cooperation. I am grateful for their blessing in that decision.

I am exceptionally grateful to Will Lippincott, my agent, who eventually beat me into shape after too many years writing for academic journals. At Oxford University Press, I thank my editor Theo Calderara, who has done wonders to the manuscript, patiently chipping away at the narrative to reveal to me a much better product that lay beneath. Thanks also to assistant editor Marcela Maxfield, who kept us wonderfully organized in the final stages, and to Lauren Hill, Phil Henderson, and Jonathan Kroberger for their work on marketing and publicity for the book. I also thank Suvesh Subramanian for his careful and thorough help in the production process. A special thanks is due to my research assistant Silvia Spodaru, who has provided fantastic research support and tracked down hard-to-find images sometimes apparently into the depths of the underworld.

Other colleagues who have immensely helped me with the ideas herein over the years include Dan Blumstein, Clark Barrett, Rob Boyd, Joseph Bulbulia, Terry Burnham, Sarah Coakley, Oliver Curry, Dan Fessler, Martie Haselton, Ferenc Jordan, Barnaby Marsh, Michael McCullough, Ryan McKay, Michael Murray, Martin Nowak, Dominic Tierney, Drew Rick-Miller, Michael Price, Zoey Reeve, Pete Richerson, Rafe Sagarin, Montserrat Soler, Robert Trivers, Paul Wason, Harvey Whitehouse, Mark Van Vugt, David Sloan Wilson, Paul Zak, and last but not least, the late Jeffery Boswall, teacher and friend, who was a committed humanist but is nevertheless one of the spirits that lives on with me.

This book would never have been conceived without the intellectual curiosity instilled in me by my parents, Roger and Jennifer Johnson. Whatever else they believed in, they always believed in me—even when good evidence may have been lacking. My sister Becci has my love and thanks for her unswerving support of my family and our adventures, and for reminding me how to keep my feet on the ground even when my head is up in the air.

Finally, no book would have been written without the love and support of my wife Gabriella de la Rosa, whose insights, help, and sacrifices over many years of this work have been superhuman, if not supernatural. She has taught me how to see the world in new ways, and what matters in our own world. I owe a great debt of gratitude and Lego time to my son Theo, and my daughter Lulu, who at age five has asked me some of the most penetrating questions about life and our place in the universe. For some questions, we still have no answers.

GOD IS WATCHING YOU

CHAPTER 1

WHY ME?

We reap what we sow.
                                    —*Proverb*

On December 26, 2004, twenty-year-old Rizal Shahputra was working at a mosque in a town called Calang, near Banda Aceh, at the
northern end of Sumatra. He did not yet know that 150 miles to the
south of Calang, a massive 9.2 "megathrust" earthquake had cracked
the seabed of the Indian Ocean, releasing gigantic landslides under the
sea and sending a series of 100-foot tsunamis bearing down on the
low-lying coasts of fourteen countries. Sumatra was the first landfall.
Rizal and his coworkers felt a jolt, and soon afterward children ran
into the mosque screaming a warning, but it was too late. Rizal was
engulfed in the tidal wave of water and swept out to sea.

The 2004 Indian Ocean Tsunami was the second largest earthquake
on record, and the longest ever—lasting around ten minutes. The whole
planet vibrated by 1cm, and other earthquakes struck as far away as
Alaska. In the ensuing hours, tens of thousands would die. And Banda
Aceh, so close to the epicenter of the quake, was one of the worst hit.
But Rizal Shahputra was alive. Clinging to floating branches, he found
himself drifting helplessly on the ocean. At first, there were others
around him as well, holding on to debris. But gradually they disappeared and he saw only bodies. Eventually he was alone. For days he
drifted, with no sight of land. At one point a ship passed, but it did not
see him and disappeared again over the horizon. Reciting verses from
the Quran when he felt hungry, and surviving on rainwater and coco-
nuts he found floating on the water, Rizal held on. On the ninth day,
chief officer Huang Wen Feng aboard the container ship *MV Durban
Bridge*, sailing out of Cape Town, saw a tiny speck on the expanse of

open water and decided to take a look through a pair of binoculars. To his surprise, he recalled, it was "a man waving frantically for help." The ship stopped and Rizal Shahputra was rescued—100 miles out to sea.

When the waters receded, over 230,000 people were dead. Hundreds of thousands more were injured. Millions had lost family, friends, homes, and livelihoods—Rizal among them. Most victims were poverty stricken citizens of poor countries, where the devastation only added to ongoing social and economic hardship. With the shock of the sheer devastation and loss of life that grabbed the world's attention for a few weeks, many people found themselves asking a simple question: Why? Why had hundreds of thousands of people suddenly lost their lives out of the blue? Why these particular people? Why those countries? Why now? And of all times, at Christmas? Of course, for Rizal Shahputra and the many millions of people around the Indian Ocean facing or following the tidal wave, the question was much more pertinent: "Why me?" Later, trying to explain his survival, Rizal said, "I believe the angels were with me ... They saved me from the tsunami."

Many of us have asked exactly the same sort of question. Living in a world of plentiful disappointment, hardship, misfortune, loss, disaster, disease, and death—as well as miraculous stories of survival against the odds—we constantly find ourselves challenged to search for meaning behind life's trials and tribulations. The same questions surface after almost every devastating personal or natural disaster. Why did this house vanish and that one stand? Why did this person live and that one die? And what had they done to deserve it? We are somehow loath to believe that misfortune should strike the innocent. It makes no sense.

WHAT GOES AROUND COMES AROUND

At first glance, the search for meaning in such events may seem like a quirk of western culture, perhaps the lingering influence of

Christianity in our cultural heritage. But growing evidence from anthropology and experimental psychology shows that finding meaning in natural events is, in fact, a universal and prominent feature across cultures. Humans the world over find themselves, consciously or subconsciously, believing that we live in a just world or a moral universe, where people are supposed to get what they deserve. And we appear to have been thinking this way for a long time. Our brains are wired such that we *cannot help* but search for meaning in the randomness of life. It is human nature.

But it's a double-edged sword. As well as believing that if we do bad things, *bad* things will happen to us, we also believe that if we do good things, *good* things will happen to us—that our good deeds will somehow be rewarded and our misdeeds somehow punished. It's as if we have a morality machine in our minds, enticing us to be good, deterring us from being bad, and keeping score. The expectation of reward and punishment is not an invention of human culture; it seems to be a fundamental element of human psychology. It dictates the way we see the world, the way we live our lives, and the way we advise others to lead theirs. Of course, when we do good or bad things we *are* rewarded and punished in tangible, material ways, by friends, colleagues, the police, or the dreaded taxman. But our brains take reward and punishment to another level. Above and beyond any material repercussions, we can't help but worry that our actions will come back to haunt us in more intangible and nonmaterial—supernatural—ways.

Such beliefs are most obvious within the major world religions, where the possibility of supernatural reward and punishment is not just a figment of people's imagination, but doctrine. Historically, for example, many Christians fully expected God to reward and punish good and bad behavior in this life, or afterward in heaven or hell. The threat is pretty explicit in the Bible–for example, in the story of the Flood, when God decided to annihilate all of humanity apart from Noah and his family and start from scratch, or Sodom and Gomorrah, which God burned to the ground with fire and brimstone for the sins of their inhabitants. Similar ideas are found in other religions too. Muslims and Jews do not believe in a Christian-style hell, but they

3

nevertheless anticipate positive consequences for good deeds and negative consequences for bad deeds.

For Hindus and Buddhists there is no all-powerful God, yet how one acts in this life will define how one is reincarnated in the next. Similar beliefs are found in small-scale societies as well and, as we shall see, are ubiquitous among modern, ancient, and indigenous religions alike. People the world over are anxious to earn favor with—and cautious not to offend—their God, gods, ancestors or spirits, or to avoid generating bad karma. There are, of course, numerous complications and exceptions among and within different religions, but they are variations on a deeper theme. The fundamental concept that anchors them together is that worldly deeds are expected to have supernatural consequences. Sooner or later, people expect payback. Consequences may be delayed or displaced, but this ambiguity is part of its effectiveness. Just because bad deeds don't seem to be punished straightaway doesn't mean they won't be later. We can never be sure we are off the hook.

HUMAN NATURE

While the idea of supernatural observation and punishment finds its most overt expression in religion, it can be found much more widely as well. The expectation of payback is something fundamental to human nature and the human brain, and that means it applies to us all—from devout religious believers, to agnostics, to avowed atheists. And even where we might least expect it. A friend of mine works in a big London investment bank. One day he went with his colleagues for a certain kind of ice cream—a Magnum bar—while waiting to hear the outcome of their first million-dollar deal. Ever since, when a deal was on the line, they were compelled to find exactly the same ice cream. When Magnum bars were hard to find, they experienced tangible panic and nervousness, a nagging fear that the deal will fall through. Something felt out of whack.

Sometimes such superstitious beliefs occur subconsciously and we are not even aware of them. But everyone, at one time or another, displays some kind of superstitious beliefs and behaviors as we go about life, much to the delight of anthropologists. Baseball players carry out careful rituals, such as tapping home plate with the bat a certain number of times or jumping over the baselines as they run on to the field. Soccer players point up to the sky when they score. President Franklin D. Roosevelt assiduously avoided having thirteen guests for dinner or travelling on the thirteenth of the month. Winston Churchill stroked black cats to get good luck. Harry S. Truman hung a horseshoe in the White House, and Admiral Lord Nelson nailed one to his mast. Jennifer Aniston must step right-footed onto a plane and tap the outside. Gun's N' Roses' singer Axl Rose would never play a concert in a town beginning with M (he thinks the letter is cursed). President Barack Obama carried a lucky poker chip during the 2008 presidential campaign. And so on. Such beliefs and rituals may seem bizarre, but they are important to people and hard to break. We may not readily notice or acknowledge them in our own lives, but they crop up all over the place in everyday activities, from wearing lucky charms to crossing our fingers to knocking on wood. There can be a powerful feeling that if one omits or changes the ritual, the universe will conspire against us.

Many of these secular superstitious beliefs seem to carry no moral content, and thus differ from religious beliefs, which generally do. But the point is that all such beliefs stem from a common underlying expectation: If I don't do what I think supernatural forces require—whatever those forces and requirements may be—I will face payback. Adding morality into the equation only seems to increase the likelihood and severity of punishment. Suddenly it has ethical valence, and the social obligation adds to its power. In countless everyday events from whispering a little prayer to putting on a lucky shirt to avoiding walking under ladders, we all find ourselves beholden to some greater force of nature that we would find hard to explain to a psychologist trying to account for our beliefs and behavior, or to an economist trying to account for our use of precious time, energy, and resources.

In disaster or crisis, the effect can become elevated to new heights. When the stakes are high, the meaning of events become all the more significant and our search for them more intense. We act as if our thoughts and actions will be judged, if not by God, then by some other cosmic, karmic, or supernatural force. And again it does not have to be religious. We find ourselves imagining what our parents, spouse, or boss would think of our thoughts and actions, even if they are miles away and will never find out. We often feel that we are being monitored. We talk of eyes burning into the backs of our heads, the walls listening; a sense that someone or something is out there, observing our every move, aware of our thoughts and intentions. In *Treasure Island*, when Jim Hawkins discovers the dead Captain Flint's secret treasure map, he suddenly begins to fear every sound and movement: "The fall of coals in the kitchen grate, the very ticking of the clock, filled us with alarms. The neighbourhood, to our ears, seemed haunted by approaching footsteps."[1] Strikingly, the feeling is intensified precisely when we do not want to be watched, such as when we are doing something selfish, self-indulgent, or wrong. At such times we cannot help feeling that even though we may be alone—or perhaps especially *because* we are alone—some kind of higher power is watching us and marking up our ledger.

We may even find supposed evidence for such supernatural activity stronger than evidence against it, particularly when such beliefs are powerfully bolstered by our cultural surroundings. Why would you *not* believe that God or ancestral spirits or karma were real if everyone else did, if your forebears had always done so before you, and if your parents, relatives, and friends had ingrained it into you since birth? Atheism is hard in a world of believers.

Indeed, most indigenous cultures do not debate the existence of God (or gods) the way we do. There is no question of whether supernatural agents exist or not. Instead, what we describe as religion is part and parcel of everyday thinking and living. There is little division between what is religious and what is nonreligious. There is no pressing search for evidence of supernatural agents as if they were a hypothesis to be tested, but rather a search for ways to live alongside them, just as you have to live alongside the cycle of the seasons or your neighbors. Indeed, the whole logic of cause and effect can become reversed: rather than John Doe's bad deeds being thought to lead to misfortunes, misfortunes that befall John Doe are interpreted as *evidence* that he must have done something wrong. A belief in supernatural consequences may not always be present or strong in all individuals, but it is a remarkably widespread and pervasive aspect of life. Christian or Hindu, New Yorker or ancient Hawaiian, devotee or atheist, we tend to lead our lives *as if* we are being watched—whether by God, spirits, ancestors, or some other ordering principle of the universe.

THE GOALS OF THE BOOK

The goals of this book are threefold. First, to show that belief in supernatural reward and punishment is no quirk of western or Christian culture. It is a *ubiquitous phenomenon of human nature* that spans cultures across the globe and every historical period, from indigenous tribal societies, to ancient civilizations, to modern world religions—and includes atheists too. Heaven and hell may be the best-known versions of supernatural reward and punishment, but they are mirrored by a panoply of others that are thought to occur in this life—notably negative outcomes such as misfortune, disease, and death—as well as in the hereafter. And while we in the West tend to think of a single, omnipotent God as our judge, in other cultures rewards and punishments may come from a pantheon of gods, angels, demons, shamans, witches, ancestors, ghosts, jinns, spirits, animals, sorcerers, and voodoo. In other cases there is no specific agent at all, but supernatural consequences still come as the result of karmic forces of nature and the universe. The variation is remarkable, but there is a clear underlying pattern: our behavior is strongly influenced by the anticipated supernatural consequences of our actions. They make us question our selfish desires, deter self-interested actions, and perform remarkable acts of generosity and altruism—even when alone and even when temptation comes knocking at our door.

7

The second goal of the book is to argue that this is no accident. Rather, it is an *evolutionary adaptation*. The ability to anticipate rewards and punishments arising from our behavior would clearly have been favored by Darwinian natural selection, because it promoted survival and reproduction. And among humans, I argue, this extended to the anticipation of *supernatural* reward and punishment. Why? Because god-fearing people were better able to avoid raising the ire of their fellow man, lowering the costs of real world sanctions, and raising the rewards of cooperation. This is not a just-so story about how humans as a whole are better off if everyone is nice to each other—nice guys fall right into the jaws of natural selection. Rather, when humans evolved the capacity for complex language and theory of mind (the ability to know what others' know), our behavior became increasingly transparent, and selfish behavior and social transgressions risked increasing costs—from retaliation or reputational damage. Avoiding these costs—an evolutionarily novel danger of life in cognitively sophisticated social groups—ushered in a new era in which the *suppression* of selfishness became a vital ingredient of an individual's evolutionary success. The looming threat of supernatural punishment deterred selfish behavior and increased cooperation, and this was a good thing for individuals as well as for society.

The third goal of the book is to think through the *implications* of all this. How has a concern for supernatural consequences affected the way human society has developed, how we live today, and how we will live in the future? If it is so important for human cooperation, might it even have increased the scope of what humanity could (and still can) achieve? Does it expand or limit the potential for local, regional, and global cooperation today? How will the current decline in religious belief (at least in many western countries) affect selfishness and society in the future? What does religion's spread in other regions, and the rise of fundamentalism, bode for the future? And what, if anything, is replacing our ancient concerns for supernatural punishment as the means to temper self-interest, deter free-riders, and promote cooperation? In short, do we still need God?

WHO CARES ABOUT THE WRATH OF GOD?

In today's world natural phenomena have well-known scientific explanations, so is there really any room left for beliefs about supernatural punishment? We could perhaps explain our payback mentality as mere ignorance. Once we understand science and statistics, there is no puzzle. Stuff happens. In fact, however, there are at least three reasons why we can't simply dismiss belief in supernatural punishment.

First, people continue to believe that events have supernatural meaning *irrespective of scientific knowledge*. Even with a PhD in plate tectonics from MIT, we might understand the physical causes of a devastating earthquake but be no less puzzled as to why it happened *on that day* rather than the day before, or why *those* people were to die rather than some others. Scientists may be better able to rationalize the underlying *mechanisms* of events, but they are not immune to the workings of their own brains, which also strive to understand the *meaning* of such events: the "Why Me?" question. Even atheists—consciously or subconsciously—continue to expect the machinations of payback, whether via some form of superstition, folklore, karma, just-world beliefs, fate, destiny, comeuppance, just desserts, luck, or misfortune. Knowing science doesn't free us from human nature. If anything, it makes the puzzle of meaning more acute—it is the last question standing.

Second, if we want to understand the deep, evolutionary origins of religious beliefs, then we must pay attention to the role of supernatural punishment *in our prescientific past*, not just its role today. Before scientific explanations emerged, things we now take for granted—the sun, moon, stars, eclipses, seasons, thunder, lightning, rain, fire, droughts, births, deaths, mental illness, disease—were more or less unfathomable. As is evident from indigenous (and many modern) societies around the world, supernatural agents are routinely assumed to be responsible for such natural phenomena. Indeed, they are called upon to maintain or alter them—for example in sun worship, rain dances, or shamanic healing. What we would call supernatural causation is just part of normal life. The situation was not much different in the West before the Enlightenment. Until fairly recently in human history, supernatural forces were automatically thought to be the cause of much of what life threw at us. It was not a theory, it was how the world worked.

Third, *several billion people on Earth do believe in supernatural reward and punishment.* Among the world population of some 7 billion people, around 4 billion are Christians, Muslims, or Hindus, and another 2 billion or so subscribe to a variety of other religions. Only around 500 to 750 million are atheists (many of the remainder are people who do not identify with official religions but have personal religious beliefs that include expectations of supernatural rewards and punishments). If we want to comprehend the roots and regulators of social cooperation in these huge majorities, understanding beliefs in supernatural reward and punishment remains essential.[2]

A NEW SCIENTIFIC PERSPECTIVE ON RELIGION

An army of scientists, philosophers, and writers—Richard Dawkins, Daniel Dennett, Sam Harris, and Christopher Hitchens among them—have set up their barricades and clamored for the expulsion of God from a world in which he no longer belongs. Not only should God be expelled, they say, because science has explained religious beliefs away as the neurological equivalent of junk DNA—an accidental byproduct of humans' big brains—but also because God and his trappings inflict untold misery, stupidity, and war on a world that would be better off without Him.

A major problem with the assault by Dawkins, Dennett, Harris, and other so-called New Atheists—and the often labeled "science *versus* religion" debate—is that it gives the impression that there is some kind of consensus among evolutionary scientists on the causes and consequences of religious beliefs. But this couldn't be further from the truth. They bludgeon past a new and rapidly growing brand of scientific research on religion that argues that religious beliefs and behaviors evolved precisely because they help us.

Although there are indeed many scientists who argue that religion is an accidental byproduct of human cognitive mechanisms that evolved for other reasons, there are many other scientists who argue that religion is the polar opposite of an evolutionary accident—rather, that it is an evolutionary *adaptation*. New work in anthropology, psychology, and evolutionary biology suggests that not only do religious beliefs and practices bring important advantages in today's world (such as promoting cooperation and collective action), but that they were actually *favored by* Darwinian natural selection because they improved the survival and reproductive success of believers in our ancestral past. This offers a scientific alternative to the Dawkins model of God-as-accident. It also offers a striking twist on the old science and religion debate: religion is not an alternative to evolution, it is a *product of* evolution.

CONSEQUENCES

When we do something selfish or wrong, even if we are alone and could never be found out, we nevertheless find it hard to shake a sense that somehow our actions are observed and disapproved of by someone or something. It's not logical. It's not rational. But it turns out that such a belief is common to religious and nonreligious people alike. In fact, it seems to be ubiquitous across history and across cultures—part of human nature. We may reject the idea of this or that god, or any official religious affiliation, but even atheists are not immune to the all-too-human feeling that our good deeds will somehow be rewarded and our misdeeds somehow punished. Children see supernatural agency all over the place and find it perfectly normal, and even years of secular education can fail to eradicate these beliefs. Atheism is a battle not just against culture, but against human nature.

Amidst the dazzling diversity of religions across the world, a few key elements stand out as common to them all. This book focuses on what I think is one of the most important, widespread, and powerful—*supernatural punishment*. The idea that one's good and bad deeds will be observed, judged, and rewarded or punished by God or some other supernatural agent is a recurring feature of virtually all of the world's religions, both past and present. That may seem surprising. But it should not be. If there were no *consequences* to following a given set of religious beliefs and practices, then why would anyone do so? Without supernatural consequences, good or bad, religion falls apart. Supernatural consequences are the fundamental framework around which other elements of religion are built. It forms the core of the machine, an engine for religion to work. And as we shall see in the next chapter, it is punishment, rather than reward, that wields the greater power over us.

CHAPTER 2

STICKS AND STONES

Bad is stronger than good.
—*Baumeister et al.*[1]

Getting around in traffic-laden cities is a constant bugbear of modern life. Wouldn't it be great if you could just jump on a bike for a few blocks, and then dump it at your destination, leaving it for someone else? A fleet of free bicycles would be a handy public good. People have repeatedly come up with such schemes over the years, but they always seemed to fail. A famous one was launched in Amsterdam in the 1960s, with a fleet of smart bikes painted white. Even in peaceable Amsterdam, sadly, the bikes gradually disappeared. Some turned up as far afield as Moscow and the United States. I grew up in Cambridge, England, where there seem to be even more bikes than books—so many that there was always a good selection to be found lying at the bottom of the river Cam. Nevertheless, Cambridge launched its own free community bike scheme in 1993. Armed with the lessons of previous attempts and the special location of a sleepy varsity town, everyone confidently expected great success. It would be a model for a future without traffic and an end to global warming. How did this one go? Not great—all 300 bikes were stolen on the first day! Today, city cycle schemes are popping up all over the place. But guess what—you have to pay, and often you have to register so that they can track you down if a bike goes missing. The idea was a good one. But it took a long time for people to realize, or believe, that you needed an effective system of payment and a threat of punishment, as well as the obvious potential rewards, before giving shiny new bikes away to strangers.[2]

DARWIN'S PUZZLE

Cooperation is so fundamental to social life that we hardly even notice it, but for decades it has posed a major puzzle for evolutionary theory. Charles Darwin worried that his entire theory of evolution by natural selection might be blown out of the water by the presence of altruism in nature (helping others at a cost to oneself). The logic of natural selection and the "survival of the fittest" would seem to leave little room for individuals who sacrifice their own precious time and energy to benefit others. Why, for example, would bees evolve to engage in suicidal stinging behavior in defense of their colonies? Why would female meerkats forego breeding in order to help rear the young of others in the clan?

Biologists after Darwin didn't worry about this too much because they tended to see his theory of evolution as working for "the good of the species" or "the good of the group," a self-regulating process in which altruism made perfect sense. An individual bee or meerkat might lose out, but all in the service of the greater good of the colony or clan. Only later was this view revealed to be fatally flawed. As biologists learned more about DNA and the mechanisms of genetic inheritance, it became clear that individually costly traits that benefited the species as a whole would quickly disappear from the gene pool. Since genetic information is only passed from parents to offspring, any strategy so foolish as to make sacrifices for unrelated others would die a quick death at the hands of natural selection.[3]

Rejection of the good-of-the-group argument, however, brought biologists back to Darwin's original puzzle: If altruism is not favored by natural selection, how can it evolve? Sadly for Darwin, this vexing problem was not solved until long after his death. The discovery of the gene as the fundamental unit of natural selection paved the way for a new understanding of altruism in nature. What seemed like altruism could be explained as a consequence of how genes, if not always the individuals that carry them, could benefit from helping others.

There are four main ways that this can work. First, there is "kin selection," in which cooperation can arise among *related* individuals because they share some of the same genes. Second is "reciprocal altruism," in which *unrelated* individuals cooperate because favors, even if costly when provided, can be returned later. Third, there is "indirect reciprocity," in which helping is favored because it leads to a good reputation that in turn attracts cooperation from others. Finally, there is "costly signaling," in which cooperation can arise because the very fact that it requires sacrifice advertises one's quality or commitment and thus wins mates or other benefits. These theories have very successfully explained pretty much all instances of cooperation among animals in nature. Although animals may sometimes appear to be acting altruistically, they are only doing so if it serves the interests of their genes.[4]

By contrast with cooperation among animals, however, cooperation *among humans* is still relatively poorly understood. Although people do increase their levels of cooperation when the four traditional theories of cooperation outlined above are in play, we nevertheless continue to cooperate even when these conditions are absent. In the words of experimental economists Ernst Fehr and Simon Gächter, "people frequently cooperate with genetically unrelated strangers, often in large groups, with people they will never meet again, and when reputation gains are small or absent." Human cooperation therefore remains an "evolutionary puzzle." A growing body of experimental work concurs that when asked to play simple games that are designed to represent everyday social dilemmas, people from both modern and preindustrial societies around the globe cooperate to a greater extent than can be accounted for by traditional evolutionary theories.[5]

Several solutions to the puzzle of why humans are so cooperative have been proposed, and which are the most important remains a topic of ongoing dispute. One method, however, that has proved to be resoundingly and consistently important in theoretical models, experiments, and in real life is punishment.

CARROTS AND STICKS

If we want someone to do something, we have two basic tools: the carrot and the stick. This book focuses on the role of the stick—*punishment*—

as a source of cooperation. This is not because it is more desirable, but simply because it turns out to be more effective.

Rewards are, of course, an important method of promoting cooperation. If you expect to be rewarded for something, you are surely more likely to do it. And in the real world, there often *are* rewards for cooperative behavior, so the incentive is tangible. It is easy to show with game theory that rational decision-makers (an important caveat, meaning people who objectively weigh up costs and benefits in making their decisions) will help each other if there are reliable positive payoffs for mutual cooperation.

It is also important to note that punishment cannot exist on its own—it is *rewards* that people are working toward in the first place. Punishments are add-ons that help to achieve those rewards. Life with punishment but no rewards would be a kind of slavery, where the only "reward" would be avoiding punishment from time to time. That would be a meaningless existence. Instead, in society we are constantly immersed in working toward rewards—whether that means public goods, food on the table, money, status, respect, family, fun, whatever. I am certainly not arguing that rewards are unimportant. In a way they are everything. But for a society to offer them, and for people to attain them effectively while dealing with noncooperators, punishment is the key that unlocks the door.

Both rewards and punishments are necessary for the system as a whole to work, but a large body of research in economics, political science, and evolutionary biology has reached a common consensus: Cooperation can be much more reliably achieved (sometimes *only* achieved) by the deterrent effect of punishment. As we will see, punishment has an intrinsic leverage on our brains and behavior, and turns out to be especially good at promoting cooperation.

THE PUNISHMENT BIAS

While rewards are important, they are not effective enough *on their own*. This is self-evident when we look at the world around us. Societies are grounded in mutual cooperation toward a common good. By working together, everyone benefits—from roads and hospitals to armies and markets. Yet every modern and many ancient societies found it necessary to develop police and courts and jails. Society doesn't work without them. Rewards are not enough, because although they may encourage *many* people to cooperate, they cannot prevent *all* people from cheating. Even if the rewards of cooperation are large and obvious to everyone, they provide no deterrent against cheats, who can make even greater gains by shirking the costs of cooperation and enjoying the benefits nevertheless. I can still drive on the highways regardless of whether I have paid my taxes.

This reflects the fundamental paradox behind the famous game called the "prisoner's dilemma." The name of the game comes from a story of two partners in crime who are arrested and brought in for questioning. They are interrogated in separate rooms, and each is told they will get a full reprieve if they testify that the other committed the crime. If only one of them does that, the tattler will get off free and their silent partner will pay the full penalty. But if neither do so, the police will not have enough evidence and they'll both get a lesser penalty. So, while it is mutually beneficial for them both to keep quiet, there is a strong incentive to rat the other out. But if the other does the same thing, then they'll both end up in the clink. So what should they do? This game has captivated scholars from a variety of related disciplines for decades, because it leads to a paradoxical outcome. Even though each player is aware of the substantial rewards if they cooperate by keeping quiet, rational actors "defect" (that is, choose not to cooperate) because defecting is the only way to avoid the nightmare scenario of exploitation by the other player, forcing you to pay the full penalty alone. Critically, there is no credible deterrent against such self-interested behavior. So, in trying to avoid the worst-case outcome, both players end up worse off than if they could only have somehow agreed to cooperate.[6]

An example of the prisoner's dilemma comes from Puccini's opera *Tosca*. The hero, Cavaradossi, is to be executed for helping an escaped political prisoner. His lover Tosca, desperate to save him, is offered

Table 2.1 The prisoner's dilemma in Puccini's opera *Tosca*. Tosca's decisions and outcomes are in *italics*. Scarpia's are in Roman type. Here, "cooperate" means to stick to one's part of the deal, and "defect" means to renege on one's part of the deal.

		Scarpia	
		Cooperate	Defect
Tosca	Cooperate	Scarpia beds Tosca but Cavaradossi lives *Tosca saves Cavaradossi but sacrifices her virtue*	Scarpia beds Tosca and kills Cavaradossi *Tosca sacrifices her virtue for nothing*
	Defect	Scarpia frees Cavaradossi and is killed *Tosca saves her lover and her virtue*	Scarpia killed *Cavaradossi killed*

a devil's bargain by the chief of police, Baron Scarpia. Scarpia says he will issue blank ammunition to the firing squad, as long as Tosca agrees to sleep with him. What should they each do? Table 2.1 lays out the various options they are faced with, and the payoffs work out nicely to make it a prisoner's dilemma (with some assumptions, such as that death is worse that infidelity; fine with economists, perhaps not with some theologians). This example highlights the key problem of the game—cooperation hinges on whether one can trust the other player to stick to the deal when it comes to decision time.

After the intermission, one returns from a glass of champagne to find that all hell has broken loose in Rome and, worse, that the economists were right all along. Tosca and Scarpia had agreed to the deal, but then *both* secretly reneged on it. Scarpia says he will issue blanks but does not, and Tosca goes to Scarpia's apartment but murders him. The outcome is, tragically, the punishment for mutual defection: Cavadarossi is shot by the firing squad, and Scarpia is killed. To top it all off, a devastated Tosca leaps from the castle roof to her death. With hindsight, of course, both Tosca and Scarpia (not to mention Cavadarossi) would have been much better off if they had stuck to the deal and at least achieved the rewards for mutual cooperation. Although they were rightfully distrustful of each other, this paved the way to disaster.

The tragedy of the game is precisely why the prisoner's dilemma and similar social dilemma games that pit self-interest against the interests of others have become such important models for understanding cooperation, and the failure to cooperate. There may be stupendous rewards waiting for people who cooperate with each other, especially if the same individuals play the game with each other many times, but when push comes to shove they often do not trust each other enough to attain or sustain cooperation. Without some form of punishment for defectors, cooperation breaks down. The prisoner's dilemma may seem to represent a rather specific and unusual scenario, but in fact it shows up all over the place in human and animal life. It turns out, for example, to underlie arms races in international relations, inaction on climate change, obstacles to trade, and even natural phenomena such as why trees grow so tall—giant redwoods could save terrific resources by only growing to, say, 50 feet (they usually grow to over 200), but in the competition for light, whoever grows that bit taller at the expense of the others will do better. In all these domains the safest option is to defect, because at least that way you avoid the worst-case scenario, which is exploitation by others. The game theory of prisoner's dilemma has been studied to death, and consistently shows that the best decision is to defect, *irrespective of what the opponent does*, because no alternative strategy can bring greater gains if the other player is also playing his best strategy (this is the so-called Nash equilibrium, after the Princeton mathematician John Nash).[7]

The prospects for cooperation under the prisoners dilemma are gloomy indeed. But the problem becomes magnified even further when one starts looking at cooperation in *groups*, rather than cooperation between two individuals. There are now more people not to trust, and greater anonymity in shirking the burdens of cooperation. The magnitude and prevalence of this challenge was powerfully articulated in American ecologist Garrett Hardin's famous "tragedy of the commons." Hardin illustrated the problem with the historical example of open pasture shared by herdsmen. "The commons" is an old English term for grazing land freely available to all. Because no one individual is responsible for the land, each herdsman has an incentive to let his cattle graze there as long as possible, but this leads to overgrazing. If grazing land were private property, the owner would be able to ration the use of the land to preserve it over the long term and prevent it from being ruined. If no one owns it, such restraint is hard to achieve or enforce. The tragedy of the commons underlies many modern examples of failure to cooperate over so-called common pool resources such as overfishing, space junk, and climate change. No one owns the sea, space, or air.[8]

The same underlying dilemma has been examined in "public goods games" by economists. Common pool resources are things we jointly *deplete*. Public goods are the reverse: things we jointly *create*. We all benefit from public goods, such as roads, police forces, and air traffic control, but someone has to pay for them. If everyone contributes a bit, everyone benefits a lot. But as with the prisoner's dilemma, the best strategy of all for an individual is not to contribute anything. Such free-riders will benefit from the roads whether they helped to create them or not. Plus, they get to save whatever resources other individuals had to invest to make them. Over time, free-riders enjoy greater payoffs than everyone else. Because of the enormous importance of overcoming this collective action problem in everything from taxes to war to climate change, it has become ever more critical to understand the conditions under which people will be more likely to cooperate to achieve public goods and preserve common resources.

To this end, economists (and more recently anthropologists, psychologists, and political scientists) have been herding people into laboratories to play public goods games under experimental conditions. In a typical experiment, subjects are seated anonymously in cubicles with computers, allocated randomly into groups, and then have to decide how much money to contribute to a common pot over several rounds of play. The highest payoff would be achieved if everyone contributed all of their money to the pot, because the pot is increased by some amount (e.g., by 40 percent), to represent the public good, and then the total is shared out to everyone. However, if only some people contribute, then you would be better off keeping your money because the total pot is shared out to everyone whether they contributed or not (so you'd be better off keeping your money *and* raking in the profits from others' contributions). Hence, the game presents a social dilemma like a multiplayer version of the prisoner's dilemma—whether you should cooperate or not depends on what others are likely to do. And since a rational player in this scenario should defect, the basic economic prediction is that people will not cooperate.

There have been numerous variations of this game, and the consistent result is that without other sorts of intervention, cooperation quickly breaks down. Although people tend to contribute some of their money at first, no one wants to be the sucker and people rapidly reduce their investments over subsequent rounds of the game. The hunt is therefore on for ways to prevent such social tragedies and promote cooperation. Figure 2.1 shows the results of a famous example of this kind of experiment

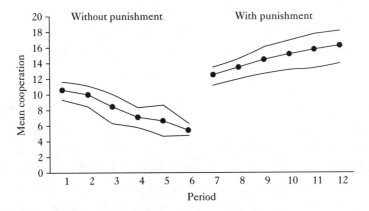

Figure 2.1 The famous result from Swiss economists Ernst Fehr and Simon Gächter's public goods game experiment. Over sequential rounds of the game (periods 1 to 6), people contributed less and less to the common pool, which is a standard result in public goods games. However, as soon as people were allowed to punish (from period 7 on), cooperation not only immediately jumped up to a high level, but then also increased further over time (periods 7 to 12). Black dots show the mean levels of cooperation each round; outer lines show the 95 percent confidence intervals. © Fehr, E. and Gächter, S., "Altruistic Punishment in Humans," *Nature* 415, no. 6868 (2002): 137–40.

conducted by Swiss economists Ernst Fehr and Simon Gächter, which included a crucial twist. After running the game as it is normally played, with people's levels of cooperation starting low and declining further over time as usual, the experimenters suddenly changed the rules. Now, free-riders could be *punished*–that is, participants in the experiment could use some of their money to impose fines on people who did not contribute. The effect was remarkable. As soon as people could punish, cooperation skyrocketed, and then actually increased over subsequent rounds of the game. Punishment solved the collective action problem.[9]

Having lived in Switzerland myself for a year, where this experiment was conducted, I was initially dubious about these results. Swiss people seem to love rules. I got a lot of practice in form-filling and rule-following, starting with registering with the police on arrival. A colleague of mine had a neighbor come over *on Christmas Day* to complain that their children were making too much noise! So maybe Fehr and Gächter's result was a quirk of Swiss students preferring the introduction of some rules to an otherwise anarchic game. But no. Since this well-known experiment was published, the same result has been replicated many times in many different countries. Swiss people may like rules, but the willingness to punish defectors seems to be a more general trait of human nature.[10]

A more serious question about these results is whether it is artificial to look at a group on its own, operating in isolation. This may not be very representative of real life. There are usually multiple groups one can join and leave. And what if there is competition between one's own and other groups? Perhaps punishment increased cooperation *within* the group, but at the cost of the group's overall success. The costs of punishment (to both the punished and the punishers) may mean the group does poorly compared with other groups. Furthermore, free-riders might be tempted to leap between groups, exploiting one and moving on before punishment can be administered. Further empirical experiments using public goods games explicitly addressed these problems. Ozgur Gurerk and his colleagues in Germany put more people in a public goods game, but this time people could choose to move between alternative groups—a group with sanctioning and a group with no sanctioning.

The results were striking. Although initially most people chose to be in the group where there was no punishment (who wouldn't?), as average contributions flagged, people started moving into the sanctioning groups by choice. Over time, more than 90 percent of people moved into the sanctioning group. And it paid. These individuals made higher contributions, and enjoyed larger returns. As the authors reported, "a sanctioning institution is the undisputed winner in a competition with a sanction-free institution." What was even more notable is that in the sanctioning institutions, people could *reward cooperators* as well as punish defectors. However, *only* negative sanctions—punishment—had a significant effect on how much people cooperated. Punishment seemed to have a special leverage.[11]

Much debate has ensued over *why* people punish, and whether and when people do so outside of laboratory experiments. But what matters for our purposes is that punishment works—it is very effective at promoting cooperation. In many contexts beyond the laboratory rewards similarly turn out to be less effective than punishments in promoting cooperation. Despite the promise of mutual benefits, cooperation collapses in real-life groups if there are no additional arrangements to prosecute or punish dissenters. Cooperation, even when it does occur, is a delicate balance. The presence of just a single or a few cheats can cause people who would otherwise be cooperative to withdraw their cooperation (as happened, for example, in the decline in cooperation over time in the Swiss experiment shown in Figure 2.1). The decline occurs because people notice others being selfish and reduce their own cooperation accordingly. No one wants to be exploited by cheats. Such results have led to a convergence of opinion among experimental economists, game theorists, and evolutionary biologists that wherever self-interest conflicts with group interests, cooperation in groups will not emerge unless there is some additional factor, such as the punishment of defectors.[12]

While economists and other scholars are deeply involved in identifying solutions to this problem, human society appears to have worked out what must be done long ago. As much as we might dislike having them and paying for them, massive and complex institutions are in place to detect, catch, and punish cheats. Most people are cooperative and do their bit. But there are always some who don't. Punishment, or at least the threat of punishment, is necessary for society to function.

The idea that punishment has a special power over and above rewards has parallels in some famous works of philosophy. For example, Adam Smith is renowned for his idea that if people simply work for their own rewards, a positive side effect of this self-interested behavior is that it helps society too (as if led by an "invisible hand"). Despite the underlying, self-interested striving for rewards that has been seized upon by many modern economists wishing to show that unregulated free markets lead ineluctably to benefits for all, Smith also said, in fact, that coercion was necessary as well to keep people in tow. Another example is Niccolo Machiavelli, who was convinced that in weighing up the relative effectiveness of punishment and reward, punishment wins hands down. In his famous advice to his benefactor, *The Prince*, he asked:

> [Is it] better to be loved than feared or feared than loved? It may be answered that one should wish to be both, but, because it is difficult to unite them in one person, it is much safer to be feared than loved, when, of the two, either must be dispensed with. . . . men have less scruple in offending one who is beloved than one who is feared, for love is preserved by the link of obligation which, owing to the baseness of men, is broken at every opportunity for their advantage; but fear preserves you by a dread of punishment which never fails.[13]

THE NEGATIVITY BIAS

So punishment seems to have an advantage in theory, but are people really influenced by it in reality? There is, in fact, reason to believe that human beings are *more sensitive* to punishments than they are to rewards. We care about them more. This is important because it suggests that even in a society or a religion that emphasizes reward rather than punishment, punishment may nevertheless weigh more heavily on people's minds.[14]

Psychologists have found that *negative* events and phenomena are much more potent influences on our thinking and behavior than positive ones. One study found, for example, that negative images elicit more attention and neural activity than positive images. In another study, subjects were able to locate a lone angry face in a grid of happy faces more quickly than the opposite—people were drawn to faces that signaled potential danger. These and many similar results were encapsulated in a review of this literature by the eminent psychologist Roy Baumeister and colleagues with a title that says it all: "Bad Is Stronger Than Good." They identified an overarching bias—a so-called negativity bias—in the way human brains work. Negative stimuli have greater effects than positive stimuli on cognition, motivation, emotion, information processing, decision-making, learning, and memory. It can be difficult to show in any one case that good and bad factors are objectively equal. However, a large number of experimental studies replicated the phenomenon in controlled laboratory conditions, where positive and negative stimuli could be carefully manipulated to be identical in frequency and magnitude (e.g., equivalent monetary reward and loss), and other explanations could be ruled out.[15]

In one prominent example, Nobel laureates Daniel Kahneman and Amos Tversky found a powerful and consistent empirical anomaly in which people are *risk-averse* when facing choices among *gains*, but *risk-prone* when facing choices among *losses*. When facing losses, people are especially eager to avoid them and willing to take risks to do so. This gives rise to the psychological phenomenon of "loss aversion," the preference for avoiding losses over acquiring equivalent gains. In other words, the cost of giving up an object is perceived to be greater than the cost of acquiring it. Our aversion to a loss of, say, $100 is greater than the increase in satisfaction from a gain of $100. The effect is a powerful one, such that people tend to value losses as around *twice* that of equivalent gains. The specter of loss looms large in the eyes of human beings. As tennis ace Jimmy Connors quipped, "I hate to lose more than I love to win." Since punishment tends to involve losing something you already have (whether freedom, health, status, reputation, resources, or whatever), it is likely to hurt more than rewards appeal.[16]

After scrutinizing a vast range of empirical and experimental studies, Baumeister and his colleagues concluded that the primacy of bad over good was so systematic and pervasive that it represents a "general principle or law of psychological phenomena." Despite efforts to identify contrary instances, the authors found "hardly any exceptions ... [and warned that psychologists may] have overlooked the full extent of its generality." Other reviews of the literature have reached a similar conclusion. The accumulation of multiple corroborating psychological phenomena, all acting in a negative direction, makes these conclusions striking.[17]

Negativity bias may seem to contradict other phenomena in psychology such as optimism and overconfidence. While there are indeed instances in which people are overly optimistic, these tend to arise in *assessments of the self*. By contrast, the negativity bias tends to arise in *assessments of others or one's situation*. We may be overoptimistic about ourselves, our control over events, and our future—three phenomena collectively known as "positive illusions"—but we are negatively disposed toward our environment and other people within it. On the whole, humans are risk-averse, highly sensitive to threats, and on constant alert for dangers in our environment. The negativity bias is significant because it means that punishment is likely to be more important than reward in explaining human behavior—what people actually do. Finally, it is perhaps no accident that of the six basic emotions that are universally displayed by all human beings across cultures—happiness, surprise, disgust, sadness, fear, and anger—only one is positive, and four are negative (surprise could be either). Negative phenomena, it seems, have long dominated our evolutionary history.[18]

SURVIVAL OR EXTINCTION

If the bias toward punishment and negativity is universal—and the preponderance of evidence suggests that it is—it stands to reason that it is the result of evolution. But why? Why should negative events have influenced natural selection more than positive ones?

The reason is simple. Genetically speaking, life is not a safety net from which organisms can take risks to strive for greater heights. Rather, life is a high wire from which organisms must do everything they can to avoid plunging to their deaths. For the entire 3.5 billion year history of life on Earth, organisms have faced lethal dangers in the guise of predation, rivals, competition, disease, starvation, weather, extreme heat and cold, natural disasters, and so on. Eking out an existence in an adverse environment was an exercise in survival.

This has, not surprisingly, given rise to a range of fundamental features of living organisms, which privilege the task of avoiding negative consequences. Most prominent among these—including among humans—are the fight or flight response, a sensitivity to threat, and an aversion to predators and strangers. We also have a fear response that is processed by the brain's amygdala, before there is even any conscious awareness of the situation. This is not just a human trait. We share the automatic fear response with all other mammals and it goes far back in our evolutionary history. In general, organisms are risk averse, develop multiple redundant systems for safety, have sophisticated immune systems, evolve armor, avoid danger, and are on high alert.[19]

Nature has come up with two broad strategies to deal with the fundamental problem of risk, known as r and K strategies. The r strategists tend to reproduce in great quantities, like spawning salmon. Given that any single offspring is statistically unlikely to survive, they swamp the local environment with thousands of young—investing in multiple tickets in the lottery, as it were. K strategists take the opposite approach. They have just a few or a single offspring, but invest great resources and time to ensure their chances of survival into adulthood. Watching over one or few individuals constantly, and providing them with food and security, they can be shielded from the vagaries of nature. Humans, along with most mammals, are K strategists. One consequence is a suite of adaptations to maximize the survival of our precious few lottery tickets—perhaps most importantly physiological and psychological mechanisms to avoid dangers at all costs. We maintain a constant vigil over infants for years and go to extraordinary lengths to protect, defend, and provision them.

A broader evolutionary perspective also suggests that extinctions reinforce the primacy of negative versus positive outcomes. In the long view of evolutionary history, certain animal groups have tended to survive and others—perhaps a billion species—have gone extinct. In addition to the mechanisms that serve to preserve life *within* generations, there are also features that make some types of species more likely to survive even rare and unpredictable occurrences such as climate change and extinction events. These are features that steel them from major disasters. The types of organisms that stick around over the long term, therefore, are the ones that have not only developed strategies to deal with the quotidian dangers of everyday life, but which have also developed traits that can withstand calamitous events. The survival of mammals and the demise of the dinosaurs in the cretaceous extinction is one prominent example.[20]

A final reason why evolution may have primed us to pay particular attention to negative events is because optimal decision-making appears to depend not just on picking the option with the greatest expected returns, but avoiding the worst costs. This has become known as error management theory. A smoke alarm is deliberately set to go off slightly too often. The cost is occasional false alarms when you burn your toast, but this is a small price to pay to make sure the alarm will not fail to go off in a real fire. We err on the side of caution. This "smoke detector" principle has analogues in a variety of decision-making problems ranging from engineering and medicine to economics and public policy. As it turns out, evolution seems to have encountered exactly the same problem (and devised similar solutions). In a given domain, if the costs of false positive errors and false negative errors have been asymmetric, then natural selection would favor a bias toward whichever was the least costly error over time. There are two conditions: (1) there must be uncertainty about the true signal (otherwise evolution or engineers would simply build a perfect device that guessed right 100 percent of the time); and (2) false positive and false negative errors must entail different costs over time (otherwise they would cancel each other out). Wherever these conditions are fulfilled, natural selection will favor the evolution of biases in decision-making.[21]

In evolution in general, including for humans, all decisions are made under some form of uncertainty, and the costs of different outcomes are unlikely to ever be identical. Human psychology is littered with cognitive biases, and it seems that most if not all of these biases arose as solutions to a common underlying problem: avoiding the worst possible mistakes. Error management theory therefore suggests that human cognition is generally organized around the probability and severity of negative (rather than positive) events in our environment.

There are deep evolutionary reasons why organisms fear—and try to avoid—negative events. Positive events are important too, but negative events have greater leverage because of the risk of injury or death. Positive events may be desirable, but they are not imperative. As the saying goes, the fox is only running for his dinner; the rabbit is running for his life. The key to survival is avoiding negative events.

WIRED FOR PUNISHMENT

There is one final reason to believe that negative events exert a special power on human beings: the way the brain works. It seems that rewards and punishments, however they may be distributed in the environment, are dealt with very differently at the neurological level. It has been suggested that reward is generally dealt with by what psychologist Jeffrey Alan Gray termed the behavioral approach system (BAS), while punishment is dealt with by the behavioral inhibition system (BIS). This has important consequences because each system has different effects. The BAS is responsible for generating emotions, whereas the BIS is responsible for threat and novelty—and for learning. So not only are reward and punishment processed differently, but they can have different impacts on future behavior. Most notably, the part of the brain that is involved with learning primarily deals with *punishment*, not reward. This suggests a powerful reason why negative events seem to play a special role in learning of all kinds, including the development of moral behavior, norm compliance, and cooperation, and even why people and organizations tend to learn more from disasters than successes.[22]

Much is changing in this field of research due to rapid advances in neuroscientific methods and knowledge. But as psychologist Timothy Wilson put it, "there is increasing evidence that positive and negative information is processed in different parts of the brain," with dramatic consequences for our thinking and behavior. These differences may extend to how much we even recognize the influence of positive and negative phenomena. Wilson explains that "there may be a division of labor in the brain, in which the unconscious is more sensitive to negative information than the conscious self." Our response to negative information and events, therefore, is to some extent beyond our awareness, let alone our control. A long tradition of research has shown that both animals and people have rapid vigilance systems—"preconscious danger detectors that size up their environment very quickly." This is the concrete neurological result of the biological imperatives that we discussed in the previous section. Information is evaluated by the sensory thalamus before it enters conscious awareness, and if the information represents a possible threat, a fear response is triggered. This is clearly adaptive, since dangers can be a matter of life and death, but opportunities cannot. One consequence of this system is that we have a natural, unconscious aversion to significant losses, even if the net payoffs from enduring them would be positive. The evolutionary, psychological, and neuroscientific evidence nicely converge here. While still under study, Wilson notes that "there is at least the possibility that the adaptive unconscious has evolved to be a sentry for negative events in our environments."[23]

These results tally with the more general idea that there are both conscious and subconscious mechanisms that process information, so-called dual-process theory. System 1 is the instinctive, fast judgment and decision-making that happens without our being aware of it. System 2 is the reflective, slower judgment and decision-making that incorporates more rational thought. Psychologists have become extremely interested in which kinds of thoughts and behaviors are associated with each system, or how those thoughts and behaviors change if they are confined to System 1 or System 2 alone. For our purposes, what is interesting is that System 1, in its encapsulation of

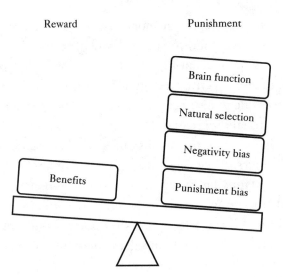

Figure 2.2 The leaning tower of punishment. Both reward and punishment are important parts of the equation of incentives for cooperation, but punishment weighs more heavily on people's shoulders, minds, evolutionary history, and neurological wiring.

fear responses and the processing of punishment, may generate a fundamental tendency to fear negative events and to learn disproportionately from them, irrespective of how we may consciously reflect on the same events. We may, therefore, deny the special influence of punishment even as it drives our behavior. Overall, we have now seen multiple reasons why punishment is more effective than reward (Figure 2.2).[24]

THE ANIMAL ORIGINS OF PUNISHMENT

Punishment is not unique to humans. Nor should we expect it to be. Much of the logic of game theory laid out above, such as the prisoner's dilemma and the collective action problem, applies to any interacting entities, whether they are humans, firms, nation-states, chimpanzees, or slime molds. The logic of evolutionary theory is very much like the logic of game theory. Different strategies have different payoffs, and those gaining the highest payoffs tend to do best. However, in regular game theory, payoffs represent money or resources, whereas in evolution, payoffs correspond to survival and reproduction. A winning strategy leads to more offspring, and thus spreads in the population at the expense of less effective strategies. Wherever a strategy of administering or avoiding punishment increases payoffs, therefore—whether in terms of material resources or Darwinian fitness—we should expect to find it in the behavior of humans and any other organism.[25]

Indeed, empirical evidence reveals that many animals punish. Biologists Tim Clutton-Brock and Geoff Parker wrote a major review of this topic in the scientific journal *Nature* in 1995 called "Punishment in Animal Societies." They pointed out that while evolutionary biologists had paid considerable attention to *positive* reciprocity (such as the reciprocal altruism we met earlier), there had been scant attention paid to *negative* reciprocity (i.e., retaliatory aggression). This was an important oversight because, they noted, retaliatory aggression–punishment–is common among social animals. Clutton-Brock and Parker gave a range of examples of animals punishing each other. Rhesus macaques, for instance, who find a food source but fail to announce this to other members of their group are punished by physical aggression. Punishment may even extend to kin. In vervet monkeys, individuals who are pushed off food sources may find and attack *relatives* of the usurper. Perhaps the most common form of punishment is the displays of aggression one sees in the dominance hierarchies of many social mammals, such as wolves or lions, where dominant individuals snarl and bite at those who overstep their rank. Punishment is part of everyday life.

But punishment can be more or less important depending on the context. Clutton-Brock and Parker identified five key situations in which punishment is likely to be particularly common: (1) dominance hierarchies (in which, as we just saw, dominant individuals aggress against subordinates); (2) theft, parasitism, and predation (in which individuals attack or harass cheats, exploiters, or predators); (3) mating bonds (in which males tend to punish noncooperating females); (4) parent/offspring conflict (in which adults punish offspring); and (5) the enforcement of cooperation (in which individuals may punish those who do not cooperate or join alliances). In a remarkable example of the latter, queens of some wasp colonies are "regularly aggressive to inactive workers, chasing, biting, grappling or bumping them." Even in the highly cooperative social insects, therefore, we can find coercion and punishment at work in maintaining the harmony of the hive.[26]

Punishment is not a fixed behavior in a given species or a given setting, however. It also varies among individuals. Some individuals may engage in punishment more or less depending on their status or stage of life. Indeed, individuals may alter their punishment strategy to reap changing benefits as they move between different contexts. For example, chimpanzees often intervene in the fights of others, but how they do so depends on their social status: "Rising males initially support winners in quarrels between other group members but switch to supporting losers as soon as they reach the alpha position." They ally with stronger sides to ride the waves to the highest rank and, once there, use punishment to maintain the status quo—siding with the weaker sides to preserve the balance of power.[27]

But does this animal aggression really count as what we call punishment? Arguably, the essence is the same. Punishment can be defined simply as wherever an individual incurs a cost to themselves in order to impose a cost on another (see Table 2.2). This is often referred to as "spite," since it is mutually costly—I take a hit to make you take a hit. Punishment is clearly costly to the punished, since they suffer the consequences. But punishment is also costly to the punisher, for a variety

Table 2.2 Behavior can be categorized by its effect on the actor (rows) and its effect on others (columns). Given positive or negative effects on each participant, this gives four broad types of behavior: mutualism, altruism, selfishness, or punishment. Punishment is defined as behavior that imposes a cost on both participants (lower right cell).

		Effect on Others	
		+	−
Effect on Actor	+	Mutualism	Selfishness
	−	Altruism	Punishment

of reasons. Key among these are: (1) time (it takes time to seek out and punish a target); (2) energy (it takes resources to carry out punishment); (3) opportunity costs (one could be doing something else with the time and energy it takes to punish); and (4) retaliation (punishing someone may make enemies or rivals, and risks injury if the victim reacts aggressively).

Punishment might appear maladaptive at first glance. It is a seemingly wasteful act that incurs costs for both parties. Wouldn't natural selection favor individuals who abstain from spiteful behavior and preserve their time and energy for more productive activity? In fact, punishment can bring net benefits, despite its initial costs, if it changes a victim's *future* behavior, or if it changes the behavior of *other* observers. If punishment serves to deter the perpetrator from repeating some undesirable behavior in the future, or deters would-be perpetrators from trying something similar, it can pay off in the long run. Punishment works because it prevents, deters, weakens, banishes, or even eliminates offenders.

This logic applies to humans as well as animals, of course. For example, it can be advantageous to go to great lengths and costs to punish someone for a minor infraction if the result was that they never did it again. Or it can be advantageous to punish someone for a minor infraction if the result is that onlookers, seeing the grievous consequences, refrain from ever trying to take advantage of *you* in the future (they might pick on someone else instead or not do it at all). This is why we talk of "talking a stand" or "making an example" of a particular infraction. Although any given instance of punishment may be costly, and may even be out of all proportion to the crime, it serves as a signal that it will not be tolerated in the future. The logic is manifested in organizations as well as individuals. Businesses are often willing to pester people endlessly and expensively for the payment of minor fees. It seems disproportionate and vindictive. But if they did not, the threat would be empty and people would stop bothering to comply. As long as occasional violations are chased up reliably, most people will pay up in order to avoid the (inevitable) punishment. As Thomas Jefferson put it to John Jay in 1785, "I think it is to our interest to punish the first insult; because an insult unpunished is the parent of many others."

Under the right circumstances, therefore, punishment—despite being costly in any given instance—makes perfect sense within the logic of rational choice and the logic of natural selection. As Clutton-Brock and Parker put it, "punishment is temporarily spiteful, in the same way that reciprocal altruism is temporarily altruistic [helping others at a cost to oneself]. However, in the longer run, punishment is a form of selfish behavior which benefits the punisher because it reduces the probability that the victim will repeat a damaging action or will refuse to perform a beneficial one." Clutton-Brock and Parker developed a game theoretical model to show that punishment not only works but, under a wide range of conditions, is in fact an evolutionarily stable strategy. That is, a winning strategy that spreads in the population and cannot be invaded by alternative strategies.[28]

CONCLUSIONS

Clearly, punishment is not unique to humans and has similar causes and consequences in humans and other animals. By changing the costs and benefits of our behavior, it tilts the balance away from selfishness and toward cooperation. The comparison with other animals, however, highlights aspects of punishment that *are* unique to humans. For one thing, punishment among humans is relatively rare, whereas for many animals it is constant. Punishment is also manifested among humans in a far more elaborate way—consider our complex institutions of laws, police, courts, fines, jails, and so on. And it deeply pervades politics, culture, and entertainment (why are so many novels and movies about cops, criminals, courts, and revenge?). The elaborateness of human punishment may reflect the difficulty of achieving effective cooperation without it. "Punish or perish," as mathematical biologist Karl Sigmund put it. In a land of fierce competition and fragile cooperation, punishment is king.[29]

Figure 2.3 Lady Justice standing above the Old Bailey courthouse in London, representing fairness with the scales of truth and justice in one hand, but wielding a sword in the other. With origins in Egyptian, Greek, and Roman deities, however impartial and reasoned the often blindfolded Lady Justice may have been, she and her institutions have always found it necessary to carry a sword. © iStock.com/georgeclerk.

If we want cooperation, we need both carrot and stick. In every society, in every part of the globe, and throughout history, the various rewards of social life have been accompanied by punishments for those who do not cooperate (Figure 2.3). These punishments are not always severe or consistent or even fairly administered, but they are there. Of course, punishment is by definition somewhat limited in its effectiveness, because it is a corrective measure that kicks in *after* something has happened—it comes too late to do anything about the initial infringement. As Woody Allen noted, "capital punishment would be more effective as a preventive measure if it were administered prior to the crime." Critical to achieving and maintaining cooperation, therefore, is to establish a credible *threat* of punishment in order to deter people in the first place. This is in a way more important than punishment itself.

Clutton-Brock and Parker's study of punishment in animals is especially interesting for the contrast they themselves drew with human societies. In small-scale societies, they note, individuals often have to punish minor violations themselves if it is to happen at all (as animals must do), but more serious violations tend to be punished by the group as a whole or by some specific institution or individuals designated for the purpose. This is something animals cannot do—they cannot plan, organize, delegate, or delay punishment. The reason humans adopt this institutional approach seems to be because it avoids the escalation of retaliation and reprisals between the individuals involved, and thus helps to maintain social stability.

If punishment is so important, then the significant degree of cooperation that is evident in society must owe its success to some mechanism that is good at detecting and punishing cheats. This is quite plain in today's world—we have intricate systems of laws, police to enforce them, and courts and jails to punish violators. But the very extensiveness of this solution poses a puzzle. If we cooperate today because of laws, police, courts, and jails, how did humans cooperate in our past when such institutions were weak or entirely absent? In recent history punishment abounded, with tyrannical kings and chiefs often using ruthless force to keep the citizenry in order. However, all such formal institutions of punishment are very recent compared with the much longer span of human history during which our extraordinarily high levels of cooperation evolved. During this time, some *other* mechanism to deter cheats must have been necessary. But what was it?

Three main solutions to this conundrum have been debated among evolutionary anthropologists. First, perhaps punishment was administered by some external institution rather than individuals. Second, perhaps punishment was not costly after all, so people were willing to do it. Or third, perhaps people who *failed* to punish were themselves punished. However, none of these three mechanisms seem satisfactory, as explained by anthropologists Joe Henrich and Rob Boyd:

While it is useful to assume institutional enforcement in modern contexts, it leaves the evolution and maintenance of punishment unexplained because at some point in the past there were no states or institutions. Furthermore, the state plays a very small role in many contemporary small-scale societies that nonetheless exhibit a great deal of cooperative behaviour. This solution avoids the problem of punishment by relocating the costs of punishment outside the problem. The second solution, instead of relocating the costs, assumes that punishment is costless. This seems unrealistic because any attempt to inflict costs on another must be accompanied by a least some tiny cost—and any non-zero cost lands both genetic evolutionary and rational choice approaches back on the horns of the original punishment dilemma. The third solution, pushing the costs of punishment out to infinity, also seems unrealistic. Do people really punish people who fail to punish other non-punishers, and do people punish people who fail to punish people, who fail to punish non-punishers of defectors and so on, ad infinitum?[30]

The puzzle therefore remains: Without institutions of law and order, and without a good incentive for people to punish each other, how could early human societies establish cooperation with a credible deterrent threat for cheats?

Here's one possibility. What if the burden of punishment could be off-loaded onto something *beyond* the group and, even better, onto something even more powerful than any human punisher could ever be? Like, say, God. A fear of *supernatural* punishment—divine retribution—rather than the limited and unreliable punishment of other mere mortals, may offer a key to the evolutionary puzzle. The specter of supernatural punishment may have helped to establish and sustain cooperation, without people needing to get their own hands dirty dishing out sanctions. And it doesn't matter if it is real or imagined. If people *believe* in supernatural punishment, then it becomes a deterrent in reality, regardless of whether the gods are out there or not.

But is such a belief widespread? And does it work?

CHAPTER 3

HAMMER OF GOD

If death were a release from everything, it would be a boon for the wicked.

—*Plato*[1]

At 4:53 p.m. local time on Tuesday, January 12, 2010, a 7.0 magnitude earthquake rocked the island of Haiti. The epicenter was just a few miles from the capital Port-au-Prince, home to three-quarters of a million people. Eye witnesses who had been in major earthquakes before said this one was different. In the past they were able to collect their thoughts and run to safe locations. This time, the quake was so sudden and so violent that there was nothing they could do before walls, ceilings, and roofs collapsed around them. When the dust settled, some 220,000 people were dead, and hundreds of thousands more injured, many of them horribly, with crushed limbs, head injuries, and infected wounds. Even those who escaped injury or death were victims: they had lost family members, friends, and livelihoods. A quarter of a million homes and thousands of business were damaged or destroyed, along with the presidential palace, the national assembly, and the cathedral. Haiti was devastated.

What seemed so heart-wrenching was that Haiti had already been laid to waste long before the earthquake. For decades the country had suffered political, economic, social, and physical turmoil. It had fallen victim to nature's fury several times in previous years, with a series of catastrophic hurricanes, floods, and landslides. Added to this were years of government incompetence and oppression by dictators such as the infamous François "Papa Doc" Duvalier. France and the United States had intervened in Haiti numerous times, especially after political unrest in 2004 which prompted a major UN peacekeeping operation, bringing thousands of US troops to the island. Poor governance, corruption, poverty, and crime remained massive problems. Haiti was the poorest country in the Western Hemisphere, and the United Nations called Haiti's slum, Cité Soleil, "the most dangerous place on Earth." When the earthquake hit, there were still 10,000 UN personnel in Haiti, with years of work ahead of them. That day, the UN headquarters in Port-au-Prince itself collapsed, killing almost a hundred United Nations staff including the Mission's Chief, Hédi Annabi.[2]

A question that came naturally to many people's mind was *why.* Why Haiti? How could a country that had been through so much already now deserve this? The randomness of the destruction was particularly hard to accept. Why did this building fall and that one stand? Why did this child live and that one die? The trials and miracles of the ensuing days only added to a burning desire to rationalize the disaster. Babies were born in makeshift tents with scant or no medical attention, to mothers whose husbands or other children had died just days before. Orphans were fighting for food and water. Few will forget the images of two year-old Redjeson Hausteen Claude's dust-covered face, pulled from the rubble after two days trapped beneath his home, or the smile that broke when he saw his mother. Or of the others, pulled lifeless from the rubble, found too late or too badly injured to make it out alive.

The 2010 Haitian earthquake was one of the worst natural disasters in recent years—indeed in living memory. It triggered immense displays of passion by ordinary citizens, organizations, and governments around the world, who committed money and help to the recovery effort. But what also stood out, and does so every time the world experiences a major natural disaster, is the question of reason and meaning. However much we understand science and plate tectonics, we still find ourselves asking *why* it happened, why it happened *then*, and why it happened to *them* (and thanking God it hadn't happened to "us"). It wasn't long before a familiar explanation started to be advanced by people around the world, from victims, onlookers, and leaders alike—these massive natural disasters were the hand of God.

In one extreme example, Pat Roberston, a Christian televangelist and former US presidential candidate, declared that Haiti deserved it, after it "swore a pact with the devil" to throw off the tyranny of France, and "ever since they have been cursed by one thing after the other." But it was not just religious radicals or fundamentalists who saw it this way. Pooja Bhatia, a fellow at the US Institute of Current World Affairs, was in Haiti at the time. In a *New York Times* op-ed the day after the earthquake, she wrote, "if God exists, he's really got it in for Haiti. Haitians think so, too. Zed, a housekeeper in my apartment complex, said God was angry at sinners around the world, but especially in Haiti. Zed said the quake had fortified her faith, and that she understood it as divine retribution." Even President Barack Obama, speaking after the quake, declared "we stand in solidarity with our neighbors to the south, knowing that but for the grace of God, there we go." Others blamed people's treatment of Haiti itself. A Haitian survivor told a reporter, "I blame man. God gave us nature, and we Haitians, and our governments, abused the land. You cannot get away without consequences." Others thought that it meant God had abandoned them. The night of the quake, "there was singing all over town: songs with lyrics like 'O Lord, keep me close to you' and 'Forgive me, Jesus.' Preachers stood atop boxes and gave impromptu sermons, reassuring their listeners in the dark: 'It seems like the Good Lord is hiding, but he's here. He's always here.'" Survivors prayed in the streets.[3]

The discussions of supernatural punishment surrounding the disaster in Haiti included many references to its voodoo religion, a tradition with roots in West Africa that links human communities to a pantheon of spirits, and is still the official state religion. After the earthquake, a variety of people—in both Haiti and elsewhere, and not just Pat Robertson—suggested that Haiti must be cursed, and this explained its horrendous run of bad luck. Ceremonies were performed to try and appease the spirits, in Haiti and also in far-off Benin in West Africa. Some Christian denominations in Haiti blamed the voodooists themselves for incurring God's wrath, and in February a group of Protestant evangelicals even attacked a Voodoo ceremony with a barrage of rocks. Voodoo scholar Wade Davis explained that "in traditional African belief, no event has a life of its own. Everything is connected in a flow of causal association. Many Haitians in their agony and sorrow will be asking deep and anguished questions: Why now? Why us? What more can a tormented nation and a people be expected to bear?" If the search for meaning was potent among onlookers comfortable in their homes thousands of miles away, it was nothing compared to the search for meaning among the victims themselves. How does a mother who loses all of her children come to terms with what has happened? Random chance offers no satisfactory explanation. Would she lose or keep her faith? Either outcome would speak volumes.[4]

Religious explanations for these natural disasters were as horrifying to many atheists and believers as they were compelling to others—examples of religious belief gone mad. However, the idea that God has a greater plan and that events on Earth unfold according to his will is commonplace. The extent to which He can cause or prevent natural events varies among traditions, religions, and individuals, of course. But for many believers, there is no soul-searching to be done. This was God's work. There are deep philosophical and theological debates on this so-called problem of evil—why a benevolent God would cause or allow innocent people, even children, to suffer and die. The implication is that He is not benevolent, not all-powerful, testing our faith, or that these people deserved to be killed or maimed. All such avenues lead us into difficult theological territory.

Luckily, however, the problem of evil is not the focus of this chapter, nor of the book. I am simply interested in whether people *believe* in supernatural punishment or not, and whether such beliefs are effective in changing people's behavior. Above all, I am interested in whether an expectation of supernatural punishment changes people's behavior *for the better*. But before we get on to what supernatural punishment might *do*, is it really such a widespread belief in the first place? Is it just an idiosyncratic phenomenon of certain modern, or western, religions? Or is it a recurrent or even universal feature among all human societies?

SUPERNATURAL PUNISHMENT IN THE MAJOR WORLD RELIGIONS

The obvious place to start looking is among the major world religions—Christianity, Judaism, Islam, Hinduism, and Buddhism. Is supernatural punishment a quirk of certain big religions, or is it a common phenomenon of them all?

On the face of it, Christianity might seem like the archetypal example of beliefs in supernatural punishment. After all, this is the religion that created and institutionalized the concepts of sin, confession, a Day of Judgment, and Hell, not to mention the strong association with acts of divine vengeance that God enacted in the Old Testament such as the Flood, the destruction of Sodom and Gomorrah, or the Battle of Jericho. The Bible certainly provides plentiful illustrations of the importance of God's punishment, even welcoming it. For example, the admonition "Serve the LORD with fear, And rejoice with trembling" (Psalms 2:11). In the Bible, of course, "fear" of God can be a metaphor for understanding or recognizing God, not fearing him in the literal sense. But fear him we well might. The wrath of God not only stands out in the Bible, it also pervades our history and culture, dominating, for example, the protestant ethic of the Anglo-Saxon industrial revolution and rise of capitalism, the founders and frontiersmen of America, church billboards along the highways, countless works of art, literature, and film, and timeless masterpeices such as the east wall of the Sistine Chapel (Figure 3.1).

While Christian sects vary considerably in their beliefs, an important theme they share is the idea of an omniscient and omnipotent God—one who is all too aware of human deeds. Someone who acts contrary to God's will can expect Him to know, and to risk some form of consequences in this life (misfortune, hardship, or loss) and—for many—in the next as well, in Hell. Many Christians believe, and indeed sometimes make a point of warning everyone else, that we face divine retribution for violating God's commandments. As the *Wycliffe Dictionary of Theology* sums it up, "it is plain from the bible that sin will be punished (Dan. 12:2; Matt. 10:15; John 5:28 ff.; Rom. 5:12 ff.,

Figure 3.1 *The Last Judgement*, Fresco on the East Wall of the Sistine Chapel, Michelangelo (1537–1541). After death, souls are sent to Purgatory to await entrance into heaven, or to be sent to hell. © Scala/Art Resource NY.

etc.).” Of course, as many people have pointed out to me as I've gone around peddling the supernatural punishment theory, the angry and punishing God of the Old Testament is very different from the one in the New Testament. That is certainly true. The God that unleashed the ten horrific plagues on Egypt and sent angels to kill all firstborn children had certainly become more subdued by the time of Jesus. However, the idea that God rewards and punishes on account of people's behavior is not lost. The way it is expressed and mediated has changed, but the underlying logic remains. While in the Old Testament "sin necessarily and inevitably involves punishment," in the New Testament " 'punishment' is not as common as 'condemnation'," but "to be condemned is sufficient. Punishment is implied." The wrath of God continues to be mentioned numerous times in the New Testament. Among nearly all traditions that follow the Christian Bible, believers who break the rules run the risk—at least—of retribution from God, or the denial of never ending bliss in Heaven. Jesus promised eternal life, but only for those who accepted the faith and followed its various prescriptions. Missing out on that would be punishment indeed.[5]

While rewards clearly play an important role (a Christian may be motivated by eternity in heaven as much as by the fear of hell), we saw in the previous chapter that punishment, in general, has a more potent influence on people's thinking and behavior. The same may be true in the context of Christianity—at least in some aspects of its development. For example, in considering Christian beliefs about the afterlife, historian Alan Bernstein pointed out that "the very proclamation of hell indicates that the defenders of religion found it necessary to balance the attraction of its promise with a threat for the 'others,' who rejected it or failed to meet its tests."[6]

Like any religion, there are many variants. And this variation is central to the debates and developments within that religion. But the variation itself is interesting for evolutionary theories of religion too—why are some concepts played up in some settings and not in others? Some Christians argue that punishment plays no role at all and that God is much better characterized by love and grace—not punishments, nor even rewards as such. Others bluntly put punishment center stage, with many Christian groups focusing their rhetoric on sin and the divine punishment that will ineluctably result. Still others, such as Calvinists, hold that everyone is predestined either to damnation or salvation, and individuals can do little about it. Clearly, the importance and role of punishment depends not so much on the *religion* you belong to, but on the *church* you belong to. This suggests a kind of ecology of religious beliefs and practices—a framework to understand geographic and temporal *variation* in religious traditions as adaptive solutions to local problems. Beliefs in supernatural punishment need not occur everywhere. They should occur when and where they help believers and not otherwise. This is important to keep in mind as we zip through many complex religions, but for now the core question remains whether beliefs in supernatural punishment are found among each of the major world religions as a whole.

Judaism shares Christianity's scriptural roots, and some of their beliefs are comparable. They have, of course, developed very differently over the centuries since. But in Judaism, too, there is a strong connection between moral behavior and supernatural consequences. There is no simple consensus on many aspects of Judaism, but one well-known set of core tenets is Maimonides's twelfth-century "thirteen principles of the faith," which includes the injunction that "God rewards good and punishes evil." God responds favorably to good people and good behaviors, and unfavorably to bad ones, with the wider theme being a familiar struggle to tamp down mankind's inherent self-interest. For orthodox Jews, the *yaytzer ha'ra* or "evil inclination," describes "the tendency to be selfish, harm others, and break rules." Both religions see in humans good motives working against evil ones. Judaism is also explicit in saying that individuals are responsible for their own actions, and the supernatural consequences that may result. Indeed, the Jewish *bar mitzvah* celebration (which means "son of the commandment"), is to mark the point at which a boy becomes responsible for his own sins. The father recites a blessing that, translated literally, frees him from the punishment due to the boy from now on.[7]

Supernatural punishment and reward are also central to Islam. Like Christians, Muslims believe in an omniscient and omnipotent God who knows your thoughts and actions and will bring you to account for them in this life or the next. Muslims follow detailed religious codes of behavior, often extremely strict ones, and share a moral obligation to "Be a community that calls for what is good, urges what is right, and forbids what is wrong." All human behavior is divided into five categories, the most important of which are "required obligations," such as prayer, alms-giving, and fasting, and "proscribed or prohibited" behaviors, such as theft, illicit sexual activity, and drinking wine. The faithful are rewarded for following these rules and punished for failing to do so. Of course, as is well known, Islamic Shari'a law is also strictly enforced by human institutions in many regions, and can involve severe physical punishment. But violations of the laws incur supernatural punishment as well. While some of the laws might seem somewhat arbitrary to outsiders, in the context in which Shari'a law originated and to which it subsequently adapted, there are strong indications that they were important for suppressing self-interest and promoting group cooperation. As Malise Ruthven put it: "The law is there both for the purpose of upholding the good of society and for helping human beings attain salvation." Military historian Richard Gabriel has even suggested that Islamic beliefs—particularly the notion of the *Ummah*, the wider Muslim community that one should value over and above oneself—were instrumental in bringing together roving bands of independent desert tribes into a highly effective, cooperative army in the wars of Islamic expansion.[8]

Hindus do not have a single, all-powerful God. They have hundreds of gods. Not surprisingly, therefore, they emphasize the action of broader supernatural forces rather than any one particular agent. When Hindus consider their fate they tend to think of it in terms of destiny rather than deities. So at first there appears to be a disconnect between supernatural agents and any possible supernatural punishment. However, in Hindu traditions all of life is wrapped up in an interconnected cosmic order, which affects how one will be reincarnated in a series of future lives. Those who do well and are good are more likely to be reincarnated in a desirable human form, whereas those who are bad are more likely to be reincarnated as an undesirable animal. While there is no concept of sin, or an all-powerful deity, or heaven or hell, in many ways Hinduism represents a stronger case for the supernatural punishment theory. There is no negotiation or fickleness with a particular God. If X happens, then Y will follow. Supernatural consequences do not necessarily stem from a supernatural *agent*. Rather, supernatural rewards and punishments stem from a more general but no less potent supernatural *agency*. On top of this, the many individual gods and deities—and their wishes and admonitions—represent an important part of the social landscape for Hindu traditions as well.[9]

For Buddhists, there are no gods at all. However, supernatural consequences for one's actions on Earth are no less significant. As with Hinduism, the question is how one will be reincarnated. This depends on one's ethical behavior in the present life. You might come back as a noble human, or you might come back as a lowly ant. But you can influence this by good and bad deeds, generating and building a reservoir of good or bad "karma" (a concept with related forms in Hinduism, Jainism, and Sikhism as well). Geshe Kelsang Gyatso illustrates the logic with a story about Ben Gungyel, a bandit who robbed travelers and homes, but later mended his ways and became a monk: "From the first thing in the morning until the last thing at night he judged his actions, trying to eliminate negative actions and practice wholesome ones. He added up his positive and negative karma daily by putting white and black stones on his table. For a positive action he put a white stone on his table, while for a negative action of body, speech, or mind he put out a black stone. If, at the end of the day, he found more black stones than white he criticized himself, shaking his left hand with his right and saying, 'You robber, don't you have to die? Do you have any choice in your next rebirth? Do you have any freedom at all when you commit so many bad actions'?"[10]

All creatures on Earth are seen as reincarnated beings from past lives. They may be former, or future, humans. Buddha himself is said to have lived 108 former lives. This belief, fundamental to practitioners of the faith, means that Buddhists and Hindus alike have strong reasons to follow the norms and taboos of their community—no one wants to be reincarnated as an ant. The need to balance good and bad in the supernatural realm therefore serves as a powerful incentive to suppress self-interested behavior and promote cooperation. Indeed, the Buddha's Middle Way was precisely a call to balance earthly pleasures with self-restraint, and thereby to reach Nirvana.

Although the notion of karma might appear to be somewhat vague, it is in fact fairly mechanical and perhaps even less easily circumvented than the watchful eye of an all-powerful God. It is embedded into the fabric of the universe. As Gyatso stresses, "the result of karma is fixed: positive karma brings happiness and negative karma brings suffering." Interestingly again, there is reason to believe that the negative plays a greater role than the positive—that supernatural *punishment* is particularly important. Both positive and negative karma are significant aspects of Buddhist teaching and practice, but there is a fundamental asymmetry: While building up positive karma can be agreeable, accumulating negative karma can be disastrous. With karma, as with money, staying out of debt is much more important than building up a fortune. Gyatso cautions that "we should work to eliminate even the smallest negative karma, regarding it as if it were deadly poison."[11]

SUPERNATURAL PUNISHMENT IN ANCIENT CIVILIZATIONS

The Greeks and Romans of antiquity are instructive examples from the ancient world because they were contemporaries of early Judaism and Christianity, and yet practiced very different forms of religion. Ultimately the Roman Empire converted to Christianity wholesale under the Emperor Constantine, but only after several centuries of bloody conflict between Romans, Jews, and Christians—and, of course, the crucifixion of Jesus of Nazareth by Roman governor Pontius Pilate.[12]

It was fundamental differences in belief that underlay much of this historical conflict. But once again we can see common threads that all groups shared. Religion was an extremely important part of Roman life and Romans invested considerable time, energy, manpower, and resources into the worship of a pantheon of gods—Jupiter, Mars, Juno, Minerva, and many others. There was no overarching God like the Christian one. But we see a familiar feature amidst the differences. The various Roman gods were seen as responsible for natural events, calamities, fortunes, harvests, fertility, warfare, and numerous other aspects of life. Indeed, when powerful political leaders wanted something to happen, they would enlist the services of the gods through priests and elaborate rituals. When commoners wanted or needed something, they would also appeal to the gods via idols, shrines, and rituals. Religion was a cornerstone of Roman society from lowly citizens to the rulers of the Empire. Julius Caesar himself, before he was Emperor, held the position of High Priest of Jupiter in Rome. Key to the social interplay between Romans and their gods was the idea that the spirits must be appeased and worshipped if one wanted to win their favor. You might carry out good deeds or bad, but the gods would know and you risked supernatural sanctions if you failed to play by their rules, or to pay the appropriate tributes. This was important to individuals but also to Roman society as a whole. Greek historian Polybius remarked on the utility of religion to the Roman state, concluding that it was "deisidaimonia (fear, awe, or respect for the supernatural) which held the state together."[13]

Greek myths read almost like modern soap operas, with the key difference being that the cast of actors include gods as well as humans. The Greeks had 14 major gods and goddesses—Aphrodite, Apollo, Ares, Artemis, Athena, Demeter, Dionysus, Hades, Hera, Herphaestus, Hermes, Persephone, Poseidon, and, last but not least, Zeus. These gods were constantly meddling in human affairs, to the point of taking sides in earthly disputes, getting emotionally involved with protégés or enemies, becoming enraged, envious, pitiful, generous—even falling in love with humans and bearing children with them. Although the gods could be fickle and unpredictable, this did not make them any less potent when humans considered what to do or how to behave. If anything, it made it even more important to placate them. It was unthinkable that one could go about life without careful thought to the reactions of the gods. Once again, the behavior of ancient Greeks was significantly influenced by the opportunity to curry favor with the gods and, most importantly, to avoid incurring their wrath. And again, there was an intrinsic asymmetry. Their favor might bring you happiness, power, and glory, but their wrath could kill.

But if we want to get a perspective on the deeper, *evolutionary* origins of religion, then we need to wind back the clock further. We need to develop a picture of what religion was like not just 2,000 years ago, but 20,000 or 200,000 years ago. We therefore cannot rely on written history, which doesn't go back nearly far enough. So how can we find out anything useful at all? There are, in fact, two ways. First, we can look at evidence that has literally been dug up out of the ground—archeological finds that tell us something about the practices, and by implication the beliefs, that human beings exhibited tens of thousands of years ago. Second, we can look at ethnographic descriptions of indigenous societies from around the world—those untainted by contact with more technologically sophisticated societies. These societies offer a vital window onto human origins, because many of them are thought to live much as we all would have done in the environment of our evolutionary past.[14]

SUPERNATURAL PUNISHMENT IN TRADITIONAL SOCIETIES

In 2007 I attended an academic conference on "The Evolution of Religion." Who goes to such an event, you may ask? Well, anyone who fancied escaping the snow of New Jersey in January, for one thing. The organizers had cunningly planned the conference in the beautiful Makaha Valley on the island of Oahu in Hawaii. Lying 2,500 miles from any other land mass, Hawaii is the most isolated island group in the world. It's amazing to see how small that distance becomes when academics have the opportunity to congregate somewhere exotic. But the traditions of ancient Hawaii also turned out to offer some striking insights into religion.

One day there was a break in the program, and my future wife and I drove up into the hills to visit a local *Heiau*—an ancient Hawaiian temple. Such sites are scattered across the Hawaiian Islands and were used for a number of purposes, including ceremonies and offerings to promote fertility, to bring rain, and to ensure victory in war. We left the car and followed a track into the hillside, winding our way through the incredibly lush greenery unique to Hawaii. The trees grew taller and the undergrowth grew darker until we were completely encased by tropical forest. No one else was there. As we approached the temple there was an eerie stillness, despite the constant chatter of insects buzzing around us, and the calls of mysterious birds echoing through the trees. As we approached we made out small thatched huts, reconstructed in the style of the ancient Hawaiians, and amongst them, totems of Hawaiian spirits. Intricately carved into heavy lumps of wood, these spirits were a tribute to what were—and still are—immensely important features of Hawaiian society. Most conspicuous were their bluntly carved faces, with ever-watchful eyes staring out from the timber. Totems always strike me as particularly disconcerting because the eyes are always looking directly at you from wherever you stand. Real people can only look in one direction, and you can tell whether they are looking at you or not. The eyes of the totems followed us everywhere we went, and I have to admit I began to feel somewhat self-conscious as I plodded around the sacred site, finding myself whispering instead of speaking. The spirits seemed to be all around us, watching.

The feeling was not novel. In ancient Hawaii the entire landscape of trees, plants, animals, sea, sky, and earth were said to be inhabited by spirits—spirits that observe people's every action, demand attention to their wishes and desires, and ultimately influence how people go about their everyday lives. Numerous objects and concepts in the local worldview possess what the Hawaiians call *akua*—variously defined as "spirit consciousness," "cognizant entity," or "sentient spirit."[15]

In addition to being surrounded by natural spirits instilled with *akua*, Hawaiians also live amidst a dense population of ancestor spirits. Like many other cultures, the dead were traditionally buried with practical items that would be needed in the afterlife, including mundane things such as food, bowls, and tools. But they do not disappear off to some distant realm. Ancestral spirits stick around and maintain a keen interest in community affairs, attentive to their family and friends and disliking their foes just as before. They might shiver with cold or get hungry just like the living, but their physical and mental powers are very different. They could flit around rapidly, be in several places at once, enjoy remarkable knowledge—of the thoughts of others (telepathy), of the presence of things in other places (clairvoyance), of the future (precognition)—and have power over matter (psychokinesis). These powers were localized, not infinite, but nevertheless surpassed those of living people and caused considerable concern. With these spirits hanging around, one had to pay careful attention to how one's thoughts and actions were likely to affect not only other people, but the spirits as well. Nothing was private. And everything had consequences: "In the Hawaiian view the world is alive, conscious, and able to be communicated with, and it has to be dealt with that way."[16]

To the first Europeans to visit the Hawaiian Islands, such beliefs were striking. Captain Cook famously tried to kidnap a Hawaiian king in order to use him as a hostage to recover his boat, and was killed by the king's bodyguards. As ever, the Hawaiians were just as concerned with the dead as they were with the living. Following the incident, a Hawaiian asked the British "when Cook might return, and what might he do to them." It was not that the Hawaiians saw Cook as a god. Rather, they believed that when anyone died they took on the form of "spirits with powers of retribution." They were worried he might come back to harm them, just as any other spirit might. In this case, since they had killed him, he might come back with a vengeance.[17]

The relationship with the spirits was not always a simplistic one, with direct supernatural punishment arriving every time some taboo or norm was broken. Often the relationship was more like a pact in which observing certain beliefs and rites ensured continued good relations with nature—the sea that provided fish, or the forest that provided fruit. The power stemmed, however, from the belief that by *not* doing what was required, the spirits would deny people sustenance or survival: "At every level of society in pre-Cook Hawai'i, examples are found of observances which either limited man's freedom of action or required him to put forth considerable effort in order to benefit nature.

These practices were undertaken as ethical obligations—man doing his part in the communal relationship."[18]

Even today, concern for spirits remains important as Hawaiians go about their everyday lives. Hawaiian real estate agents are "legally bound to reveal everything that might affect the value of a property, including ghosts." And around the dinner table, while people talk and joke, no malicious remarks are allowed out of respect for the *poi* on the table—since it represents the visible presence of the *kalo* plant that grew from the remains of the first Hawaiian. If people begin to gossip, the *poi* bowl is covered.[19]

Hawaiian scholar Michael Kioni Dudley conveys this remarkable worldview with a vivid example. Imagine a person opening a door and finding either a packed lecture hall or a storeroom. A westerner's reaction would be completely different in each case—embarrassment at disturbing the lecture, but no special reaction to the storeroom. A Hawaiian, by contrast, would in both cases be met by numerous sentient beings—whether people in rows of seats, or jars and cans arrayed along shelves: "For the Hawaiian, there are no empty storerooms. Confronting the world about him, he experiences conscious beings at every turn, and along with this their interpersonal demands." Simply going about their everyday lives amongst the trees, plants, animals, rocks, and ancestors was for Hawaiians a walk through a densely populated spirit world, in which people had a "constant sense of religious encounter."[20]

Supernatural consequences were just as important in ancient Hawaii as they were in ancient civilizations and in the major world religions today. But are ancient Hawaiians unusual among indigenous societies in their concern for spirits lurking at every turn? Not at all. When we look at societies in other parts of the world we find very different religious traditions, but remarkably similar organizing principles. At the western end of the Pacific, the people of Yap in the Caroline Islands of Micronesia also believe that ancestral spirits observe and react to people's deeds, and social transgressions can bear high costs. Incest, for example, is punishable by death—not carried out by any person, however, but by ancestral ghosts. This threat is by no means taken

lightly. It is widely believed that one of the offending pair, or sometimes some other innocent person in the group, will be killed, within
about two months. They might die from illness, or in an accident.
But somehow or other they will die as a direct result of supernatural
agency. The only recourse is to perform a series of rituals to find an
ally in the ghost world, and strike a bargain with them to convince
the other ghosts not to kill. But this is not always effective. The fear
of supernatural punishment for such social transgressions is significant enough that the Yapese say that incest is basically impractical
because of the risk of death. A wide range of other aspects of social
life are under the watchful eye of the ancestors, and people's everyday
behavior is significantly altered so as not to offend them. As described
by the anthropologist David Schneider, who conducted fieldwork on
Yap in the late 1940s, "it is the spirits who, long ago, established the
social and moral regulations which govern Yap life, who originally for-
bade incest, and who take action against it when it occurs. The spirits
are therefore seen as the source of morality and the locus of ultimate
authority." Supernatural punishment is not just a theological possibility, but permeates people's lives and changes their behavior.[21]

The trend continues as we look beyond the Pacific. In sub-Saharan
Africa there are strong moral codes of ethical behavior, many of which
are held sacred. And while traditional beliefs vary, a common theme is
the role of supernatural agents—particularly divinities, ancestors, and
spirits—intervening in people's lives and responding to their actions.
Here too, supernatural agents are concerned with social conduct, and
traditional beliefs "mostly involve supernatural punishment for any
behavior that contradicts the moral code of the community." Even now
that many Africans are officially affiliated with Christianity or Islam,
these beliefs are often held alongside traditional beliefs as well, and
both may have consequences for violations of local codes. John Mbiti's
extensive field research in Africa found that "any breach of this code
of behaviour is considered evil, wrong or bad, for it is an injury or
destruction to the accepted social order and peace. It must be punished by the corporate community of both the living and the departed,
and God may also inflict punishment and bring about justice." Nature

and ancestral spirits wield the power to affect weather, crops, disease, fortunes, and well-being. A German team that conducted a study of cooperation in Burkina Faso made no bones about the strictness of the system: "People believe spirits and ancestors enforce the moral code ... Supernatural forces punish anyone who violates the moral code, whether [via] severe illnesses, accidents or death." While spirits may bring both good and bad consequences, it is the expectation of supernatural punishment that is particularly central and important. The authors found that subtly priming people with traditional beliefs about supernatural punishment increased levels of cooperation in playing a standard economic game called the trust game. Fittingly, the authors noted, "Burkina Faso" means "the land of honest men."[22]

What about other regions of the world? Even cases that initially seem to have no such beliefs turn out to have remarkably analogous concepts. When Darwin visited the southernmost extreme of South America, Tierra del Fuego, he famously said of the people there that they had no religion. But that was as judged by his modern, western sense of the term. Missionaries subsequently described the "religion" of this area in great detail, and familiar patterns emerged. Gods feature as well as spirits, and the focus again is on social behavior and significant supernatural consequences—including supernatural punishment—for people's actions. Fuegan religions envision a supreme being, "an invisible, lonely being who rules all the world ... [and] grants life or death and provides for man's sustenance by apportioning food ... and he has established the tribal ethics." People are aware of this and petition him with prayers: "Sometimes they are worded as indignant and accusing complaints concerning hunger, storm, sickness, and death, and at other times they contain humble requests for calm weather, daily food, good hunting, and good health."[23]

While all these examples are striking for their many unique characteristics, the expectation of supernatural reward and punishment appears to be a common theme across major world religions, ancient religions, and the religions of indigenous societies. It is also notable that supernatural punishment tends to be expected as a consequence of behavior that breaks group norms or taboos—typically some form of selfish or antisocial behavior. Despite the diversity of beliefs, and variation in how supernatural punishment is incurred, the cross-cultural recurrence of supernatural policing calls for an explanation. The Yap example is especially intriguing because an underlying biological logic stands out: Incest has been selected against in human evolution because offspring of closely related people have a high incidence of lethal genetic birth defects. Yap beliefs are therefore likely to promote Darwinian fitness. From an evolutionary perspective, therefore, supernatural punishment of detrimental behavior such as incest is a highly adaptive belief. In fact, it is noticeable that religious norms and taboos in general tend to be closely linked to events critical for survival and reproduction: Beliefs, practices, and rituals are typically about food, hunting, sex, fertility, illness, death, rain, droughts, crops, social behavior, warfare, and group cohesion. Religion appears to be closely intertwined with human biology, behavior, and ecology and—as a consequence—evolution.

IS SUPERNATURAL PUNISHMENT UNIVERSAL?

The examples given so far cover a wide range of eras and regions, but are still only a subset of all the world's past and present religions. Do we have any reason to believe that belief in supernatural consequences, and supernatural punishment in particular, play a systematic role among all religions?

Broad surveys suggest that negative supernatural consequences are indeed common to religions across the world's cultures and throughout history. An influential study by anthropologist Guy Swanson in the 1950s looked at a sample of fifty indigenous societies around the world. Swanson found that 92 percent of these societies had at least one type of supernatural agent (or agency) that directly impacted people's lives, whether via moralizing "high gods," "active ancestral spirits," reincarnation, or some other form of supernatural sanctions affecting people's health, fortunes, or afterlife.[24]

Another cross-cultural study by George Peter Murdock in the 1970s looked at a bigger sample of 186 preindustrial societies drawn from all regions of the globe, and specifically asked how these societies that lacked modern medicine explained the causes of illness. Amazingly, he found that *all of them* attributed illness to supernatural causes of some form or other. Sometimes these arise from malicious magic by shamans or enemies, rather than the work of a spirit or god, but even in those cases the act is often retaliation for some prior deed. In effect, disease was universally seen as a form of supernatural punishment.[25]

Since Swanson and Murdock's time, further research has strongly corroborated the universality of supernatural beliefs and the anticipation of supernatural consequences for people's actions. Anthropologist Donald Brown catalogued the common features of all known societies around the world—what he called "human universals." Among the sixty-seven unique to humans, he found "belief in supernatural/religion," "beliefs about fortune and misfortune," "rituals," "death rituals," "taboos," "moral sentiments," and distinctions between "good and bad" and "right and wrong." Oxford anthropologist Harvey Whitehouse has spent his life studying different religions around the world, and has identified twelve characteristics that tend to be found among all religions, irrespective of period, continent, or culture. Included in this list are "beings with special powers," "moral obligation," an "afterlife," and—not least—supernatural "punishment and reward," all of which point to the universal importance of people's concern for the supernatural consequences of their actions.[26]

A recent analysis by anthropologist Chris Boehm looked at just eighteen societies, but all ones that were particularly good models of late-Pleistocene society—that is, societies that would have been very much like our ancestors' during human evolution. He found that *all of them* had some form of supernatural sanctions "to enforce local moral codes." Twelve stated that supernatural punishment was important "in general," and sixteen cited specific offenses for which people expect to be directly punished by supernatural agents. All of these offenses were antisocial ones—behaviors that were selfish or detrimental for the community, such as theft or murder.

The many variations in how these beliefs are manifested across the world and across history are large and interesting, but this has perhaps led some scholars to overemphasize the *differences* in religions across cultures and downplay the commonalities. When immersing ourselves in the dazzling variety of the world's religions, we have to be careful that we don't fall into the trap of failing to see the forest for the trees. The commonalities are equally striking, if not more so. All of the world's religions have, first, some kind of supernatural agency and, second, believe they are capable of inflicting supernatural sanctions on the living and often also on the dead.

Whatever its exact source and characteristics, supernatural punishment seems to be a ubiquitous, even universal belief. Indeed, the fact that so many means (types of agents and punishments) have been devised with the same end (supernatural consequences for people's actions) strongly suggests convergence on an effective solution to some common problem. Supernatural punishment—including punitive costs as well as withholding benefits—may have been instrumental to the functioning of human societies.

The reason we are interested in supernatural punishment is its potential for promoting cooperation. We have seen that it is widespread but how might it relate, specifically, to cooperative ventures in these societies? In a 2005 study called "God's Punishment and Public Goods," I went back to Murdock's data on 186 preindustrial societies to look explicitly at whether there was any empirical evidence that societies with gods that are more moralizing are also more cooperative. Moralizing gods were defined, following Swanson and others' previous research in anthropology, as spiritual beings who are present and active in human affairs, and specifically supportive of human morality. Across the whole sample, moralizing gods were significantly more frequent among societies that were larger, centrally sanctioned, policed, use and loan money, and pay taxes. Since moralizing gods are more likely to threaten negative consequences for those who violate social norms, these data support the idea that cooperation is aided by belief in supernatural punishment.[27]

Some of the best and most detailed studies of cooperation among indigenous societies have emerged from Nobel laureate Elinor Ostrom's research team at the University of Indiana. She travelled the world to study how people cooperate in sharing precious sources of food, water, or land—so called common pool resources. Ostrom's work led to a remarkable realization: Indigenous people often do this better than anyone else. In many well-studied examples, the imposition of western ways of managing resources actually reduced or ruined previously successful systems of cooperation—such as the intricate system of irrigation channels in the rice terraces of Bali, which run through multiple people's land. Despite technological, logistic, and economic sophistication, western models often could not match the simple but powerful local management practices. But did religion have anything to do with it?[28]

A recent study built on Ostrom's work by looking specifically at the role of religion in forty-eight cases of common pool resource management from across the world. Remarkably, there was only a single case in which people invoked purely secular means for enforcing rules. By contrast, a full thirty-nine of them believed there were supernatural sanctions in store for those who violated social norms surrounding the protection and use of the resource. For example, the Shona of Zimbabwe "believe that a snake with special powers guards the sacred forest and that trespassers who see it may get lost, may become insane or may even die." Teasing apart the types of sanctions across all forty-eight societies revealed that twenty-seven of them relied on supernatural sanctions *alone*, twelve relied on both supernatural and secular sanctions, and just one relied on secular sanctions alone (the remaining eight were unclear or lacked enough data to tell). The study also looked at what *kind* of supernatural agents are responsible for punishments. Supernatural punishments were variously ascribed to the work of gods, spirits of nature, ancestors, or some combination of the three, and "it was not uncommon for groups to believe that multiple kinds of supernatural entities inhabited resources and enforced rules."[29]

Another interesting finding was that *supernatural* agents were predominantly believed to exact *material* punishments. Of the thirty groups for which there was enough information to judge, only one was primarily concerned about "immaterial sanctions"—such as social or spiritual consequences in an afterlife. All the other societies cited direct, "material" punishments in this life such as disease, death, misfortune, crop failure, pests, bad weather, natural disasters, and a range of other earthly events (in sixteen cases they worried *only* about material punishments, and in thirteen cases they worried about both material and immaterial punishments). Although supernatural agents may be ethereal, the consequences of crossing them were tangible indeed. The most common expected penalties were also far from trivial: disease came out top, featuring as a threat in nineteen of the societies, followed by death in sixteen cases, and various unspecified misfortunes in fifteen. As the authors of the study concluded, "when religion is activated in the management of natural resources supernatural monitoring and punishment are often integral parts." It was punishment for infringements, rather than reward for compliance, that was the primary motivating force. This strongly suggests that beliefs in supernatural punishment are a key driver of behavior—and not just any old behavior, but specifically the denial of selfish temptation and the promotion of cooperation to preserve and share group resources that are critical to survival and reproduction.[30]

Wherever we look, whether in our own religious traditions, across different cultures around the world, in ancient civilizations, or among small-scale indigenous societies that provide a window onto our evolutionary past, we find a common theme: People fear supernatural punishment for their behavior—especially when they are selfish, commit social transgressions, or violate group norms. This strongly hints at a useful, adaptive function for religion in promoting cooperation.

SUPERNATURAL PUNISHMENT TODAY

Some people—atheists and believers alike—see a fear of supernatural punishment as a quirk of religious fanatics or an outmoded historical phenomenon. In fact, however, beliefs in supernatural punishment remain alive and well. A 2007 poll by the Pew Forum on Religion and Public Life found that fully 92 percent of the US population (whether they were religious or not) "believe in God or a universal spirit"— including 70 percent of people *who declared no religious affiliation.* And popular beliefs continue to be concerned with supernatural intervention in people's lives—via both rewards and punishments. For example, 58 percent pray every day (as do 21 percent of the unaffiliated), 74 percent believe in life after death (as do 48 percent of the unaffiliated), 74 percent believe "there is a heaven, where people who have led good lives are eternally rewarded" (as do 41 percent of the unaffiliated), and despite it being downplayed in modern theology and many churches, 59 percent of people still believe that "there is a hell, where people who have led bad lives and die without being sorry are eternally punished" (as do 30 percent of people with no religious affiliation).[31]

A 2005 Baylor Religion Survey specifically asked Americans about the *kind* of God they believed in, and found that 31 percent believed in an "authoritarian" God, who is "very judgmental and engaged." Another 25 percent believed in a "benevolent" God (not judgmental but engaged), 23 percent in a "distant" God (neither), and 16 percent in a "critical" God (judgmental but not engaged). In this survey, 82 percent of respondents said they believed in heaven absolutely or probably, as did 71 percent for hell. A different question asked about whether various words described God well or not. Large majorities thought God was loving, forgiving, and kind—clearly the overwhelming characterization. Yet while 52 percent thought that "punishing" did not describe God very well or at all, 34 percent still thought it described him somewhat well or very well (the remainder were undecided). God can be loving and kind, but many people continue to believe he is punishing too.[32]

Beyond the United States, supernatural punishment also remains a widespread and potent belief in cultures around the world. A Zogby International poll in 2003 asked thousands of people who belonged to a variety of religious denominations and countries whether they "will suffer negative consequences if they disobey their religion." Answering "yes" were 95 percent of Muslims (in India and Saudi Arabia), 80 percent of Hindus (in India), 80 percent of Catholics (in Peru), 60 percent of Catholics (in the United States), 60 percent of Christians (in South Korea), and 60 percent of born-again Christians (in the United States). But what would the global figures look like if we asked a sample of *all* people, rather than just those already affiliated to a religious group?

The World Values Survey periodically conducts massive global surveys on a range of topics about personal life, economics, politics, and religion. One question they ask is whether people believe in heaven or hell. Data from the latest wave conducted from 2010 to 2014, which included a minimum of a thousand people in each of dozens of countries, revealed that 60 percent of people believe in hell, and only 35 percent do not (the remainder didn't know, didn't answer, or were not asked this specific question). The latest wave has not asked about heaven, or other supernatural agents, but in the 1999/2000 poll, 58 percent expressed a belief in heaven, and 68 percent believed more generally in supernatural forces "somewhat or absolutely" (in that wave belief in hell stood at 50 percent). These figures are remarkable, given that the samples included people whether they were religious or not. Whatever you might think about supernatural punishment yourself, it clearly remains important for significant numbers of people around the world.[33]

But does it have any real effect on people's behavior? Some large sample studies suggest it might. An analysis of the World Values Survey data found that moral transgressions were less acceptable to respondents who expressed a belief in an afterlife (heaven and/or hell), implying that people's propensity for beliefs in supernatural reward and punishment corresponds to their views on—and therefore perhaps also practice of—selfish and cooperative behavior. In another study comparing countries around the world, psychologists Azim Shariff and Mijke Rhemtulla found that the more a nation's population believes in hell, relative to belief in heaven, the lower the crime rates. Here, it was specifically beliefs in *negative* supernatural consequences that were statistically significant. Curiously, greater belief in heaven (over and above belief in hell), was associated with *more* crime. Such studies are at risk of many potentially confounding factors, but even controlling for several such variables, the results held up. Indeed, belief in hell was a *better* predictor of national crime rates than previously implicated socioeconomic factors such as GDP or income inequality. Religion, and specifically beliefs in supernatural punishment, were able to explain a large amount of variation in these data.[34]

SUPERNATURAL PUNISHMENT IN THE LABORATORY

The findings above are intriguing, but they are only correlational. Perhaps both beliefs and cooperation are driven up or down by some other, as yet unrecognized factor. To address this, one would need to take people into the laboratory and test whether experimentally manipulating beliefs in supernatural punishment affected their behavior under controlled conditions, so that causality can be examined and confounding factors can be ruled out. This might seem an unlikely enterprise, but in fact it is precisely what a spate of new studies have done.

In a study entitled "God is Watching You," Azim Shariff and psychologist colleague Ara Norenzayan had people play a simple economic game called the "dictator game." In this game, one person is given $10, and asked to decide if and how to split the money with an anonymous person. The rational response—that is, if people were solely motivated by self-interest—is to keep all the money for oneself. It is entirely anonymous and there are no consequences, so why give any of the money away? In fact, however, few people really respond that way. Numerous dictator game studies have found that people tend to give away about half the money. So much for *Homo economicus*, the hypothetical human driven by rationality and self-interest that economists have clung to for decades. The game is interesting for our purposes because it offers an established framework to see whether subjects primed with supernatural beliefs are more generous than usual. Indeed they were. Shariff and Norenzayan found that subjects gave significantly more money away if they had to first unscramble a series of words such as "spirit," "divine," "God," "sacred," and "prophet"—a technique to subtly prime people with supernatural agent concepts prior to participating in the main experiment.[35]

This was a groundbreaking study, showing that something as intangible as supernatural beliefs may be a serious promoter of cooperative behavior. But it raised more questions—two in particular. First, the experiment did not differentiate between positive and negative incentives. Does supernatural *punishment* have a special power over supernatural *reward*? In the original experiment, we could not tell. However, Shariff and Norenzayan designed a follow-up experiment to address exactly this question. This time they explicitly compared people's conceptualization of God as primarily mean or primarily generous. Using another experimental game, they found that selfish behavior was significantly less likely among people who saw God as "punishing rather than loving." Fittingly, they gave the study the title, "Mean Gods Make Good People." Other studies have also found that people have a subconscious tendency to attribute knowledge of "ill deeds rather than good deeds" to God. The idea that God knows what we have done *wrong*, in particular, seems to have a special salience.[36]

The second issue arising from Shariff and Norenzayan's original experiment was that cooperation also increased with *non*religious primes associated with secular laws and police. Hence, we were left wondering whether *supernatural* punishment is any better than equivalent *secular* punishment. In later chapters we will explore in more detail whether and why supernatural punishment may yield special advantages over secular systems of punishment—both in our evolutionary past and in modern societies. But some new experimental evidence also gets at this question. Psychologist Ryan McKay and his colleagues found that priming people with supernatural beliefs made them more likely to personally carry out sanctions against other people who were behaving unfairly, and in this case supernatural primes were indeed more effective than secular primes.[37]

Along these two dimensions, therefore—secular versus supernatural consequences, and positive versus negative consequences—there

Table 3.1 Strength of motivation for cooperation depending on the valence (positive or negative) and the source (secular or supernatural) of the consequences.

		Valence	
		Reward	Punishment
Source	Secular	Weak	Medium
	Supernatural	Medium	Strong

are both theoretical and empirical reasons to believe that cooperation is most powerfully enforced when agents are both *supernatural* and *punishing* (Table 3.1).

One concern might be that these laboratory studies are conducted in mainly western, college populations, and in cultures with mono-theistic religions. However, similar effects have been found in other populations too. In an experimental study in Burkina Faso, subjects were found to be more generous in a trust game when primed with supernatural punishment beliefs (in the context of local religious tradi-tions). The trust game works like this: player A receives some amount of money, and can either keep it, or pass some portion of it to player B. If it is passed to player B, the amount triples. Player B can then keep all of this tripled amount, or return some or all of it to player A. The game is interesting because the highest payoffs can be gained by giving everything away to player B. But that only works for both players if player B returns the trust by sending some of the money back again. Both sides, therefore, have an opportunity to demonstrate trust rather than self-interest. In the Burkina Faso study, subjects primed with supernatural punishment were 20 percent more likely to send the entire amount (among player As), and on average, returned 17 percent more (among player Bs). The authors concluded that subjects' behavior was "driven by the combination of prevailing sharing norms and the belief in supernatural punishment whenever these norms are violated."[38]

Laboratory tests may seem artificial because they isolate people from their social context, anonymize interactions, and examine tasks that are far removed from the kinds of things that people do in real life. However, this is precisely why laboratory experiments are so useful. The ability to isolate variables, control for confounding factors, and compare treatment groups allows important advances toward identifying whether certain beliefs are indeed causally related to behavioral outcomes, rather than being mere correlations. One can then use this to supplement, not replace, data from the real world. For our purposes, this series of experimental studies suggests three important things: (1) conscious or subconscious perceptions of supernatural agency increases cooperation; (2) supernatural punishment, rather than reward, drives the relationship; and (3) supernatural punishment may be more powerful than secular punishment. These are three remarkable phenomena that we will explore further in later chapters.

CONCLUSIONS

While religious traditions vary enormously, we find supernatural punishment to be a pervasive and powerful theme across the globe and across history—among modern world religions, ancient civilizations, indigenous societies, and the general population alike. And what we observed in the archives and in the field seems to hold up in the lab. Supernatural punishment beliefs are definitely out there, as a trait common to all cultures. Even more tantalizingly, several studies suggest significant links between these pervasive beliefs in supernatural punishment and people's cooperative behavior. Indeed, all human societies seem to have hit on the utility of religion in getting people to set aside self-interest and work towards shared goals. But what makes supernatural punishment so good at promoting cooperation? What is it that's so great about God?

CHAPTER 4

GOD IS GREAT

And do not fear those who kill the body but cannot kill the soul.
Rather fear him who can destroy both soul and body in hell.

—*Matthew 10:28*

After centuries of enslavement, the Israelites finally escaped from Egypt after God brought ten plagues upon the land. First he turned all the water into blood, then he brought successive waves of frogs, gnats, flies, disease, boils, storms of hail and fire, locusts, and darkness, and finally he struck down all firstborn children. At last, the Pharaoh relented, and let the people go. When Pharaoh pursued them across the desert, Moses led the Israelites through the parted waters of the Red Sea and Pharaoh and his army were drowned behind them.

The show of God's power against the Pharoah was great indeed. But it was not reserved for the enemies of his people. It soon became clear that it would be equally formidable, or worse, for any of his own that should disobey him. When Moses descended from Mount Sinai to tell the people of Israel the Ten Commandments, they were petrified. They had witnessed the power of God, descending on the mountain in fire, to the sound of thunder and lightning, the mountain shaking violently and smoking. A trumpet blast grew to a crescendo and God spoke to Moses in thunder. They begged Moses to act as messenger: "Do not let God speak to us, or we will die." Moses replied, "do not be afraid; for God has come only to test you and to put the fear of him upon you so that you do not sin."[1]

In the book of Leviticus, God laid out a host of explicit rules and rituals that, along with the Ten Commandments, the people must follow. While the blessings bestowed upon them if they followed the rules were prodigious, the punishments in store for those who broke them were horrendous. The "rewards for obedience" were attractive indeed. Through Moses, God told the people that "if you follow my statutes and keep my commandments and observe them faithfully," he would bring them rain, plentiful crops, fruiting trees, food, and security, and they would be free from fear. Old food would have to be cleared away to make way for the new. Enemies would be pursued and defeated, and the people would be "fruitful and multiply." God would "walk among you."

"But," God warned, these rewards were not left to work on their own. They were followed by a terrifying series of warnings—the "penalties for disobedience." God explained that "if you will not obey me, and do not observe all these commandments, if you spurn my statutes, and abhor my ordinances, so that you will not observe all my commandments, and you break my covenant, I in turn will do this to you":

I will bring terror on you; consumption and fever that waste the eyes and cause life to pine away. You shall sow your seed in vain, for your enemies shall eat it. I will set my face against you, and you shall be struck down by your enemies; your foes shall rule over you, and you shall flee though no one pursues you. And if in spite of this you will not obey me, I will continue to punish you sevenfold for your sins. I will break your proud glory, and I will make your sky like iron and your earth like copper. Your strength shall be spent to no purpose: your land shall not yield its produce, and the trees of the land shall not yield their fruit.

If you continue hostile to me, and will not obey me, I will continue to plague you sevenfold for your sins. I will let loose wild animals against you, and they shall bereave you of your children and destroy your livestock; they shall make you few in number, and your roads shall be deserted.

If in spite of these punishments you have not turned back to me, but continue hostile to me, then I too will continue hostile to you: I myself will strike you sevenfold for your sins. I will bring the sword against you, executing vengeance for the covenant; and if you withdraw within your cities, I will send pestilence among you, and you shall be delivered into enemy hands. When I break your staff of bread, ten women shall bake your bread in a single oven, and they shall dole out your bread by weight; and though you eat, you shall not be satisfied.

But if, despite this, you disobey me, and continue hostile to me, I will continue hostile to you in fury; I in turn will punish you myself seven-fold for your sins. You shall eat the flesh of your sons, and you shall eat the flesh of your daughters. I will destroy your high places and cut down your incense altars; I will heap your carcasses on the carcasses of your idols. I will abhor you. I will lay your cities waste, will make your sanctuaries desolate, and I will not smell your pleasing odors. I will devastate the land, so that your enemies who come to settle in it shall be appalled at it. And you I will scatter among the nations, and I will unsheathe the sword against you; your land shall be a desolation, and your cities a waste.

Then the land shall enjoy its sabbath years as long as it lies desolate, while you are in the land of your enemies; then the land shall rest, and enjoy its sabbath years. As long as it lies desolate, it shall have the rest it did not have on your sabbaths when you were living on it. And as for those of you who survive, I will send faintness into their hearts in the lands of their enemies; the sound of a driven leaf shall put them to flight, and they shall flee as one flees from the sword, and they shall fall though no one pursues. They shall stumble over one another, as if to escape a sword, though no one pursues; and you shall have no power to stand against your enemies. You shall perish among the nations, and the land of your enemies shall devour you. And those of you who sur-vive shall languish in the land of your enemies because of their iniqui-ties; also they shall languish because of the iniquities of their ancestors.

After some further details, the passage ends abruptly: "These are the statutes and ordinances and laws that the Lord established between himself and the people of Israel on Mount Sinai through Moses." This is, of course, an extreme example of supernatural punishment—for many it is not representative of God or indeed other gods, is misinter-preted, or no longer relevant. It is certainly rare among societies for a god to be quite so formidable and for punishments to be laid out so explicitly. But what the story of the ten plagues and the ten commandments does serve to do is press home the point that there are kinds and degrees of punishment that gods can threaten but mere humans cannot.

WHY GOD IS GREAT ... AT LEAST FOR GAME THEORY

Supernatural punishment is not of particular interest to many modern anthropologists, religious studies scholars, or theologians. However, it has attracted considerable attention among evolutionary researchers because of its direct implications for understanding the evolution of cooperation. As we saw in Chapter 2, there are significant obstacles to achieving successful cooperation, not least the ever-present problem of cheats and free-riders. Punishment proves to be an effective solution. But *supernatural* punishment offers a solution that may be even better still.

In its ideal form—an all-knowing, all-powerful God—supernatural punishment solves several tricky problems in the game theory of cooperation, and has intrinsic advantages over secular alternatives (Table 4.1). First, cheats are *automatically detected*. If God is omniscient, then people simply cannot get away with anything, because He is always watching and deeds cannot be hidden. Second, cheats are *automatically punished*. If God is omnipotent, then people cannot

Table 4.1 Some advantages of supernatural punishment over secular punishment in solving the problems of cooperation.

Problem	Supernatural punishment	Secular punishment
Detection of free-riders	Free, automatic	Costly, difficult
Punishment of free-riders	Free, unlimited	Costly, limited
Second-order free-riders	Absent	Present
Reprisals	Impossible	Possible
First-order free-riders	Less	More

escape the consequences of their actions, because He is always able to respond in one way or another. Third, God solves the especially vexing problem of *second-order free-riders* (people who contribute to the public good, but shirk the job of policing others). If He does the punishing, then other human beings do not have to. Fourth, there are *no reprisals* against punishers, and thus no cycles of revenge or social disruption. Supernatural agents can punish people at will but people cannot attack supernatural agents. Fifth, the formidable risks of supernatural punishment means there are fewer *first-order* free-riders to begin with, setting a higher baseline for cooperation and reducing the burden of any real-world monitoring and enforcement that human beings must do themselves.[2]

From this theoretical perspective, supernatural punishment by an omnipresent, omniscient, and omnipotent God has many powerful advantages over human punishment. Some empirical evidence bears out this claim. For example, comparisons of communal societies in the United States and Israel have found that the beliefs of religious groups were *more effective* at suppressing selfish behavior and promoting cooperation than similar but nonreligious beliefs in other groups. And even when secular institutions have emerged to maintain social order in the growth of human civilizations, evidence suggests that religion had a competitive edge, remaining a vital ingredient of social cohesion over and above legal or political institutions. While people may fear the law, the fear of God is in another league.[3]

It is worth thinking further here about just why supernatural punishment can be so powerful and real-world punishment so weak. First, God offers remarkably penetrating detection. We might worry about the watchful eyes of fellow humans, but sometimes we are alone or others are not looking. Supernatural agents, though variable in their power and characteristics across and within cultures, often have the ability to be in many places at one time, to observe people's actions, and even to have access to their thoughts. In the ideal case, cheats are inevitably caught, sometimes even before they do anything. As the Bible suggests, "whosoever looketh on a woman to lust after her hath committed adultery with her already in his heart" (Matt. 5:28). No human can match the detection abilities of God, and God is always on duty.

Second, God offers remarkably severe punishment. We might fear the wrath of fellow humans, but how bad can the consequences be? Real-world punishments may deter some, but they are by no means strong enough or salient enough to deter all. Even extreme punishments such as death are finite compared to what God can threaten—as Matthew reminds us in this chapter's opening quote. The thing about God is that supernatural punishment can be significantly worse than any earthly punishments that humans could inflict: They are not left to chance, possibly worse than death, and can last for eternity.

Of course, an omnipresent, omniscient, and omnipotent god *is* only an ideal. In indigenous and ancient societies supernatural agents typically do not have the full complement of unlimited powers that a monotheistic, Abrahamic God has. But the logic remains even if there is variance in the capabilities of supernatural agents and indeed in the extent to which people believe in them—supernatural agents still wield superhuman powers. The key point is that as the perceived probability or severity of punishment for selfish actions increases, selfish actions become an increasingly dangerous prospect and people will be more likely to avoid them. Supernatural agents are not the only way of deterring people from selfishness and promoting cooperation. But they have awesome qualities that make them especially powerful compared to human alternatives.

SUPERNATURAL PUNISHMENT AND THE EVOLUTION OF COOPERATION

Beliefs in supernatural punishment, then, might have been a powerful force in the evolution of cooperation. The basic idea is that supernatural agents work like a Big Brother looking over our shoulder, ever watchful, figures of both fear and awe that suppress our self-interest and make us more cooperative and productive. Just like the *Big Brother* reality TV show, God—or other supernatural agents—can be envisaged as a camera following us around, observing and recording everything we do. Wherever we are, whatever we are doing, with friends or strangers or alone, the camera is rolling. With Big Brother hovering over us, we are compelled to think and behave differently. Contestants in the TV show *Big Brother* are not only concerned about themselves and how they look to their housemates, they are also concerned about another audience—one that lies in another world beyond the gates of their temporary internment. And one that passes judgment on whether they will win great fortunes or be cast out of their community forever. Just like the *Big Brother* house, people in the real world live their lives uncertain about what lies beyond it, but with a strong sense that their conduct in this world will matter in the next.

Across the world and across history, supernatural agents have commonly been seen as the source of social norms and taboos, and people feel obligated to cooperate because of the threat of retaliation by these agents if they do not—that is, supernatural punishment exacted on them either in life or in the afterlife. Whether supernatural punishment is real or not is immaterial. As long as people fear it then we may expect them to modify their behavior accordingly. This follows the so-called Thomas theorem: "If men define situations as real, they are real in their consequences." Once such beliefs are established, the costs of punishment are—in theory—partly offloaded onto a supernatural actor, offering a novel solution to the puzzle of cooperation.[4]

Clearly, supernatural punishment is limited by certain bounds of credibility. As anthropologist David Schneider put it, "a supernatural sanction which specifies that the criminal's left arm will fall off at high noon on the third day following the crime cannot be maintained for long." Many alternatives *are* credible, however, and no less severe: Disease and death are commonly perceived as forms of supernatural punishment, and represent inevitable occurrences at some point or other. People generally do not expect supernatural punishment to immediately follow a transgression, and indeed often explicitly imagine that it will come much later—even after death. This makes supernatural punishment hard to falsify, adding to its power.[5]

The fear of supernatural punishment—whether there is really any God or gods out there or not—offers a powerful deterrent that may have suppressed selfishness and promoted cooperation in our past, as it still does for billions of people today. It may prevent us from doing the things we'd *really* like to do, but in the long term it can help us in a potentially more important way—saving us from incurring the wrath of our fellow man and boosting the dividends from cooperation. When taunted by desire, Big Brother keeps us moral. And that can be good for navigating our genes through the minefield of human social life and into the next generation.

It's easy to dismiss the importance of supernatural punishment because many people do not take it seriously—including quite a few religious believers and theologians, but especially the (typically atheist) social scientists and many liberal citizens of western democracies. But its deterrent effect should not be underestimated in our historical and evolutionary past. Everyday natural phenomena—the sun, moon, stars, seasons, lightning, thunder, eclipses, rain, fire, droughts, births, deaths, disease—would have been amazing, frightening, and perplexing. Until science came along, supernatural forces were the most plausible cause. Everyone else believed in them. In the absence of any better explanation, why wouldn't you?

THE TAXONOMY OF PUNISHMENT

Up to this point we have lumped various forms of supernatural punishment together and concentrated on identifying whether—as a whole—they are prominent and influential. But supernatural punishment takes many forms and has many sources. First, supernatural punishment can occur in an *afterlife*—a purely metaphysical concept that relies entirely on faith and no evidence (at least from an atheist's perspective). Second, supernatural punishment can occur in *this life*—again a metaphysical concept but one that can be seemingly corroborated by real-world "evidence" (at least from a believer's perspective), if observed worldly misfortunes are attributed to supernatural causes. Thirdly, there are different *agents* of supernatural punishment. A single monotheistic god is a statistical outlier in the full dataset of religions throughout human history. What are the differences and commonalities among the various kinds of supernatural agents and agencies across cultures? More generally, do all of these different causes and consequences of supernatural punishment have equal power in deterring selfishness and promoting cooperation? Does their diversity or combination reinforce the effect? And are they tailored to specific challenges of cooperation in a given society? By exploring the varieties of supernatural punishment, we will be better able to appreciate its utility and power among human societies.

PUNISHMENT IN AN AFTERLIFE

The idea that we may be punished after death is a fundamental concept in Christianity, with its division of the afterlife into heaven and hell. One might therefore think that afterlife beliefs are a fairly recent and western concept, and other religions that have the veneer of similar heaven and hell beliefs merely reflect the influence or lenses of western culture.

In fact, this does not seem to be the case. Belief in an afterlife is widespread around the world, and can be traced back many tens of thousands of years. Ancient burial sites demonstrate that even early humans were buried *with* specific objects. These objects are often rare, valuable, or took great skill and time to make, implying that people were thought to need them in an afterlife.

While archaeologists debate the earliest dates and different possible interpretations of "grave goods," it is clear that the practice is at least several tens of thousands of years old. Perhaps the oldest burial site discovered to date is a set of 100,000- to 130,000-year-old human skeletons found in a cave in Mount Carmel, Israel. Human remains were found with animal bones or antlers placed in their hands or tucked under an arm, along with stones and red ochre.[6] Other archaeological sites reveal that by around 50,000 years ago, humans conducted ceremonial burials in which they left the dead with jewelry, ornaments, tools, and flowers. In these later cases, archeologists themselves have argued that the practice may represent the origins of belief in an after-life. It is remarkable that by examining mere objects in the ground, we can begin to understand what was going on in our ancient ances-tors' minds. Indeed, grave goods for an afterlife may represent one of the earliest pieces of evidence for any human belief. Although we will never know exactly how long ago concepts of the afterlife emerged, the related belief that what people did in this life could have consequences in the next may be very ancient indeed.[7]

Once we get into the domain of recorded history, explicit concep-tions of the afterlife quickly appear. In ancient Egypt, mummified bodies had all of their organs removed and preserved, except one: the heart. This was so that it could be weighed by the god Thoth after

Figure 4.1 A human heart being weighed after death. The jackal-headed Anubis operates the scales, while ibis-headed Thoth, the scribe of the Gods, records the result. If it is found to weigh less than a feather, the deceased may pass into the afterlife. If it weighs more than the feather, the victim is devoured by Ammit, a beast composed of lion, hippopotamus, and crocodile. © The Trustees of the British Museum. All rights reserved.

death (Figure 4.1). If it was found to weigh more than a feather on the other side of the scales, the hapless victim passes not into the afterlife but into the jaws of Ammit, a monster part lion, part hippopotamus, and part crocodile. The Egyptian *Book of the Dead* was a collection of spells to help a deceased person pass successfully into the afterlife. Such invocations had been written on tomb walls or sarcophagi since the third millennium B.C., but an actual book written on papyrus became a common inclusion in the coffins or tombs of the New Kingdom period beginning around a millennium and a half B.C. But we don't need to rely solely on these written clues to be convinced of the importance of the afterlife to the ancient Egyptians. Standing before the massive pyramids of Egypt, which are of course giant tombs, one recognizes not only a signal of power of the royalty buried within, but also a remarkable veneration for the afterlife. The Great Pyramid at Giza has a narrow tunnel leading directly from the King's tomb to the starless portion of the sky believed to represent the gateway to the heavens, and the step structure of pyramids themselves is thought to have represented a stairway for the soul. Egyptians from all levels of society were buried with items they needed for the afterlife, from bowls and food for the lowly, to priceless jewelry and golden artifacts for the elite. At least by the time of the ancient Egyptians, the afterlife was an established, institutionalized, and vital concern.[8]

Afterlife beliefs were present in the classical world as well. Military funerals sometimes have a riderless horse accompanying the convoy. This comes from an ancient Roman custom, in which a dead soldier's horse was led behind his coffin in the procession. At the cemetery the soldier was buried, and then the horse killed and buried with him. This was not simply a tribute, but an important investment so that he would be able to ride into battle in the afterlife. Afterlife beliefs may not have been typical in Roman religion, and the greater fear was incurring the wrath of the gods in this life. But the notion was present.

The Greeks, for their part, did have explicit concepts of the afterlife, envisioning an underworld ruled by Hades where people go after they die. They did not appear to greatly fear this underworld—as with the Romans, their primary concern, rather evident from Greek mythology, was avoiding the anger of the Gods in this life. Nevertheless, there were a variety of possible consequences in the afterlife as well, with specific geographic regions of the underworld reserved for people depending on their conduct on Earth and their favor or disfavor with the gods. One of them is the Fields of Punishment, where people that had transgressed against the gods would find themselves tormented by Hades, who designed punishments tailored to their crimes. If one had done nothing too bad or too good, one could expect to end up in Asphodel Meadows, which was okay, but mundane. It was a gloomy place where people had little identity or joy. But people who had led lives of virtue or heroism would find themselves on the celebrated Elysian Fields, where they would enjoy a harmonious existence without toil. One perk of the Elysian Fields was the option to choose to be reborn again. The jackpot in the Greek afterlife was to be reborn three times, and all three times achieve entry to the Elysian Fields. If one accomplished this, one would gain elite access to the Isle of the Blessed, to live in eternal paradise.

What about the rest of the world? The concept of an afterlife, and good and bad versions of it that depend on one's worldly conduct, is actually quite common in eastern religions as well, but it revolves around reincarnation rather than any kind of heaven or hell. For example, in Hinduism, Yama, the God of Death, is assisted by Chitragupta, who "keeps detailed records of every human being and on their death decides how they are to be reincarnated." In Chinese traditions, there are the Ten Kings of Hell who "preside over the successive spheres through which a soul must pass on its way to rebirth." And among Buddhists, the wheel of life represents the "poisons of ignorance, attachment, and aversion," which one must avoid lest negative karma lead one to the "six realms of suffering" and lowly reincarnation. While precisely what happens after death is conceived of in many different ways, the concept of some kind of afterlife is common among both ancient civilizations and the major world religions.[9]

What about indigenous societies? Donald Brown found beliefs about death, fear of death, and death rituals among his list of human universals, and Harvey Whitehouse identified afterlife beliefs as one of the twelve "cross-culturally recurrent repertoires" found across the world's indigenous societies. The variation in content and importance, however, is immense. Here, let us narrow our focus to the Americas to get a more detailed sense of some of the commonalities and contrasts in afterlife beliefs. Swedish anthropologist Ake Hultkrantz spent his life studying Native American religions, and developed a remarkable knowledge of the diversity and common threads across both North and South America. Conscious of the concerns about western projections, Hultkrantz asserted that "the belief in a life hereafter was in earlier times firmly established in primitive Indian tribes and statements to the contrary should be considered with reservation." In any case, it is not so much western notions of heaven and hell that may be reflected in Amerindian beliefs. We also find elements more similar to those found in eastern religions. For example, Hultkrantz found beliefs in reincarnation to be "widespread all over America," and particularly prominent among the Eskimo. Afterlife beliefs were clearly present in one form or another.[10]

What is less clear cut is how concepts of the afterlife are related to reward and punishment for one's conduct in this life. Social status, profession, or other factors beyond people's moral conduct could contribute to determining their fortunes after death. Even so, it is not uncommon to find beliefs in which " 'the evil' are separated from 'the good.' " And again, having mentioned the Christian-sounding "good" and "evil," Hultkrantz argues that "it is doubtful that Christian impulses have stimulated the development of such a concept; rather it may be safely maintained that they have gradually come to strengthen an already established dualism." Hultkrantz stresses that a recurrent feature in Native American religions is that the afterlife is conceived of as just another version of this life, and therefore one in which social rules and consequences continue to apply: "Elimination of socially inferior elements—thieves, adulterers, murderers, and others, according to the prevailing system of moral values—is required of the living by the moral code of the tribe, and is also reproduced in the other world." "Sinners" do not go to a hell rather than a heaven, as such, but they face social consequences in whatever place they end up, and may even risk being "excluded from participation in the other world." If so, "they must rove about as ghosts on earth or else they perish on the journey to the next world or are restricted to a land different from that of the ordinary dead." Native Americans also do not have Christian-like concepts of a last judgment after death, nor of divine retribution. Yet links between conduct and consequences are evident and they often have a clearly moralizing or deterrent dimension: "If at times the destiny of the morally deficient is described in gloomy colours it is meant as a warning against antisocial tendencies."[11]

Native Americans do have strong beliefs about the soul, but they conceive of it differently than western traditions do. In many Native American religions not only are people believed to have souls, but to have *two*. One (or sometimes more) "bodily souls that grant life, movement, and consciousness to the body," and another "dream" or "free soul" which is more or less the same as the physical person themselves, but exists outside of the body. This is why sleep is respected: "When the body lies passive and immobile in sleep or unconsciousness this latter soul sets out to visit faraway places, even the land of the dead." It can thus be dangerous to wake someone when they are sleeping. Unlike shamans, who may deliberately set out to the land of the dead and return in service of some ritual or inquest, if ordinary people's free souls stray there they may become trapped and never come back. If this happens, their other, bodily soul becomes extinguished too. The picture can be further complicated in various ways. In some cases, for example, one soul goes to a kind of heaven and the other goes to an underworld. But more typically the free soul goes to the realm of the dead and the body soul dies, or may reappear as a ghost. Interestingly, it is often the very *journey* to the land of the dead that attracts concern, with various "nightmarish obstacles along the way." And here moral conduct can play a role. Among the Zuni of New Mexico, for example, those who break religious taboos risk failing to find their way to the afterlife.[12] Across Native North American religions this pattern is repeated in different forms:

There are descriptions of water and sheets of fire blocking the road, rocks threatening to crush the traveller, monsters threatening to devour him; by the gates to the land of death a human-like Cerberos is waiting to snatch his brain, thereby erasing his memory of the glory of life. Whoever happens to eat the tempting giant strawberries on the way has forever deprived himself of the possibility of returning to life. In eastern parts of North America, for example among the Ojibwa in the north and the Choctaw in the south, there is a tale about the slippery pinewood log leading over a rapid stream, across which the soul must pass; at times the log changes to a writhing snake. Balancing on the log, evil humans slip and fall into the whirling waters and are transformed into fishes and toads.[13]

In South America the details differ but key themes remain. Sometimes the medicine man travels with the deceased person to the land of the dead. The Manacica say that "the road leads through jungles, across mountains, seas, rivers, swamps, and up to the large borderline river that separates the land of the living from that of the dead. Here they must pass across a bridge guarded by a divinity."[14]

Some variations are linked to physical or ecological features of the environment. For example, the Milky Way, seen from the northern hemisphere, is divided into two different streaks. Several North American Indian tribes suggest that they represent "different passageways to the other world and of dissimilar fates after death. Tradition has it that one road leads to heaven, the other to the underworld, or that one path leads to the blessed land of the dead and the other brings downfall and annihilation." There are also curious links between afterlife beliefs and a society's method of subsistence. For example, hunting tribes tend to "recognize a heaven and a world of man on earth; ideas of an underground world are very vague." By contrast, among cultivating tribes, there are variations but "all developed from the basic pattern heaven-earth-underworld." Hultkrantz notes that this distinction extends to other regions of the world as well: "This threefold pattern is common to most archaic societies affected by the so-called Neolithic Revolution (about 10,000 B.C. in the Near East)."[15]

The fact that afterlife beliefs vary depending on a society's socio-ecological context raises the intriguing possibility that the pattern may reflect some function—perhaps an adaptive response to different ways in which society must be organized, or cooperation achieved, in order to pursue different modes of living. Hultkrantz's observation is striking because it suggests that as societies became larger, settled, and agricultural—an environment in which cooperation was harder—the emergence of beliefs in an underworld may have provided a useful additional deterrent for free-riders. A recent cross-cultural study by anthropologists Hervey Peoples and Frank Marlowe found support for a related phenomenon. Moralizing gods (rather than an afterlife per se) were more frequent among societies that relied on agriculture rather than foraging for their subsistence. They suggest that "belief in moral High Gods was fostered by emerging leaders in societies dependent on resources that were difficult to manage and defend without group cooperation."[16]

After reviewing the considerable variation in afterlife beliefs across the Americas, Hultkrantz concludes that concepts about the realm of the dead are "at times vaguely defined ... But in general, belief in the realm of the dead is there, although unformulated and without fixed outlines." And while depictions of the afterlife may often resemble the lands of the living, they are not inconsequential because they "give a picture of life on earth made better or worse." The location of this other world also varies. While it may be immediately beyond the village, as among the Cubeo in the Amazon, or a day's journey away, as among the Blackfoot, it is "more common to imagine the dwelling place of the dead as situated a great distance from this world and preferably on a different level: westward out in the sea (tribes along the Pacific Coast), beyond the sunset (Amazon tribes), in the sky (numerous hunting-and-gathering peoples) and underground (some agricultural groups, for example, the Pueblo Indians)." It is also interesting that "although the opposite can be found, it is generally true that the 'other worlds' above ground seem more propitious and cheerful than the one below."[17]

In sum, belief in an afterlife is certainly not restricted to western, Abrahamic religions. It is common among eastern, archaic, ancient, and indigenous societies as well. Despite remarkable variations, common themes include the concept of a soul, and different kinds of existence after death that are related to one's conduct in life, especially with respect to anti-social behavior within the community. As Plato noted at the beginning of the last chapter, death is no release for the wicked.

PUNISHMENT IN THIS LIFE

Whether or not people envisage an afterlife or expect it to be worse for sinners, supernatural rewards and punishments can also, of course, manifest themselves *in this life*. In some ways, belief in this-worldly supernatural punishment may be more important and more powerful than belief in the afterlife. First, real-world misfortune can be attributed to supernatural agency, representing a kind of "evidence" that may seem more compelling than whatever may lie in store after death, which no one has seen. Second, real-world punishment may also be more immediately concerning to people than consequences that lie far in the future—and for which they might have time to redress the balance. Third, real-world punishment can reduce or eliminate reproductive success, whereas punishment in the afterlife cannot (not a problem for theology, but an important one for evolutionary theories of religion).

In the West, it was not so long ago that supernatural agency was thought to be all around. As Sam Harris elaborated, "it was perfectly apparent that disease could be inflicted by demons and black magic. There are accounts of frail, old women charged with killing able-bodied men and breaking the necks of their horses—actions which they were made to confess under torture—and few people, it seems, found such accusations implausible." If mere witches could cause such misfortunes, then God must have seemed formidable indeed. Across ancient civilizations and contemporary world religions, from the Romans and Greeks to Judaism and Islam, gods have been assumed to be responsible for people's fortunes and misfortunes. And once again the pattern extends to indigenous cultures.[18]

Brown's human universals included beliefs about the supernatural and fortune and misfortune. Whitehouse's recurrent characteristics of indigenous cultures included supernatural punishment and reward. Other cross-cultural studies such as Swanson's and Boehm's also found supernatural sanctions and punishments were widely believed to be exacted on the living. Among many societies, this-worldly supernatural punishment seems to play a more important role than the afterlife beliefs we explored earlier. Hartberg's study of indigenous beliefs surrounding common pool resources found "material" supernatural punishments to be much more common than "immaterial" punishments—such as in an afterlife. John Mbiti explored afterlife beliefs in Africa, and, while they can be found, traditional beliefs are generally much more concerned with supernatural consequences for the living. Hultkrantz noted that in general, "for the American Indians as well as for most other 'primitive' peoples religion primarily serves the present life: it protects livelihood, health and success."[19]

Types of punishment vary considerably, although notably they tend to affect activities related to survival or reproduction—such as health, food, and fertility. A strikingly common one is the affliction of disease. Murdock's cross-cultural study found that supernatural agents of some kind or another were *universally* held responsible for illness. And Hultkrantz noted that diseases were far from being regarded as randomly occurring, but rather attributed to transgressions of taboos, for example among the Eskimo, Athabascan, Ge and Tupi, and the American "high cultures" (the Aztec, Inca, and Maya). An interesting facet of this-worldly punishment, as compared to punishment in the afterlife, is that by anticipating or observing such consequences one can actively seek intervention. In the case of Native American Indians, "the disease is often abolished after the patient's 'confession' of the taboo offense to the medicine man." Of course, such interventions did not always work, leaving supernatural punishment a potent deterrent. And even where they apparently did, survivors would be left with a greater appreciation of the danger of risking the wrath of the gods.[20]

Sometimes punishment in this life and punishment in the afterlife can occur simultaneously. For example, among the Yahgan, the Tzeltal, and the Dakota, "disease is thought to be a consequence of *ancestral* sins. In other words, the curse of the taboo transgression stretches over generations." Invoking the ire of supernatural agents not only against oneself, but also on fellow tribe members and kin—living, future, and dead—would be a potent deterrent indeed. And we would then become concerned about others' behavior as well as our own—yet still for our own sake. Such beliefs would be a remarkable boost for social cooperation.[21]

Clearly, both this-worldly and other-worldly supernatural punishments are present and important across societies. Any religious group that does not believe in supernatural punishment would represent an anomaly rather than the norm. Furthermore, while the types and sources of supernatural punishment vary, what is striking is that they all tend to be means to the same end: the reduction of self-interested behavior and the promotion of cooperation.

AGENTS OF PUNISHMENT

Discussions about supernatural reward and punishment often revolve around God, but of course there are many alternative agents wielding supernatural power. They may take the form of a single monotheistic God, a pantheon of different gods, angels, demons, ancestors, ghosts, spirits of nature, animal spirits, witches, sorcerers, jinns, and so on. In many traditional societies, ancestors and spirits are the most important. Some kind of supreme being is often held to have been involved in creating humans or the Earth, or creating other gods, but then to have become rather detached from people's lives and is not always regarded as central. More local and familiar supernatural agents tend to play a greater role in the quotidian concerns of life.

Ancestors can be a particularly pervasive influence. In Africa, for example, dead ancestors often continue to be regarded as members of the village. People's relationships with them thus remain important. Like living people, ancestral spirits may be family, friends, or enemies, but whoever they are they have to be treated carefully to please or appease them. They are both revered and held at arm's length. In the Americas too, "the dead occasion both fear and love but not worship." In indigenous cultures around the world, this tight communal relationship means that people are often concerned with spirits just as much as, or more than, they are with other people. Psychologist Matt Rossano examined this phenomenon in detail and found that ancestors are not just ghosts hanging around that come up in conversation or ceremonies from time to time. On the contrary, they are "ever-watchful, active players in the social world with interests, concerns, and goals that must be considered in the everyday affairs of the living. This is emphasized by the fact that the living often regard the ancestors with the same emotional cross-currents that typically characterize relations with a parent or higher-ranking associate—respect, fear, affection, and occasionally resentment ... Though physically departed, the ancestors remain active, attentive 'elders' of the earthly family." Some scholars have been obliged to treat ancestors as a separate age group in their study populations, because it is not possible to understand these societies without accounting for their relationships with the dead.[22]

Once again, the details of such beliefs vary greatly across societies, but "the notion of ancestors as 'interested parties' who desire propitiation and punish cultural transgressions is a common theme worldwide," from Africa to America, and from the Pacific Islands to Asia. Dead ancestors among the Kwaio of the Solomon Islands, for example, are thought to have a constant presence as bearers of both good and bad fortune, influencing living people's health, prosperity, fertility, and success. If norms of the community are broken, sacrifices must be made to appease the ancestors and avoid retribution. In many aboriginal Australian groups and some in Asia, the dead are particularly feared. The Sora of India believe that when someone dies their spirit becomes unpredictable and causes chaos in the community, often bringing illness or misfortune. Shamans act as intermediaries to resolve conflicts and restore good relationships, and may settle problems by assigning a troublesome ancestor to become a guardian spirit for a child given their name.[23]

But among indigenous societies supernatural agents are by no means restricted to ancestors. People have found themselves in a world teeming with spirits of other sorts as well. Like ancestors and gods, "spirits of nature"—which can be animal, vegetable, or mineral—also tend to "serve as an extension of the human social world," and their "interests and concerns must be considered in the activities of the living." We saw some striking examples in the case of ancient Hawaii in the previous chapter.[24]

Importantly, however, supernatural reward and punishment can sometimes be wholly divorced from any agent at all. In the West we are typically attuned to thinking about specific beings—a certain god or spirit, a he or a she, a human-like *agent*, who wields power. However, this is not always the source of supernatural *agency*. Eastern religions, as we saw, have a broader view of a cosmic order. There are supernatural consequences, but they derive from impersonal concepts such as karma, as among Buddhists and Hindus, not only deities or spirits. There are examples of this among traditional societies too. For example, the Iroquois speak of a "pantheistic omnipotence" that they call *orenda*: "Whereas other peoples easily regard the supernatural power as personified, the Iroqouis do not." Perhaps of greatest significance is that most societies have multiple agents, or agencies, that threaten supernatural consequences. People often face not a single supernatural agent, but a horde of them—adding to their power, domains of interest, and deterrence.[25]

THE PRIMACY OF SUPERNATURAL PUNISHMENT

Some supernatural agents are not there to deal out punishments at all, but rather to help people—often to steer them away from behavior that might anger other gods in the first place. For example, among the Pueblo Santa Ana, "each individual possesses from birth a guardian spirit who may even prevent his protégé from committing a wrong action and who receives food sacrifices before each meal." In Hawaii, ancestral gods sometimes adopt material form to help out, appearing as a rainbow, hawk, or fish, sometimes as a warning, but also sometimes as a sign of reassurance or good luck. Certain rocks have supernatural powers and can grant fertility. In Africa, people "consult with witch doctors to solve personal problems and spend significant shares of income on charms to protect against illnesses and mystic attacks." And across cultures, among the other things they do, "ancestors may simply be regarded as powerful spirits who can provide protection, fertility, and prosperity in return for honor and sacrifice." Among Native Americans there are specific cult organizations through which individuals can extract supernatural rewards. For example, one cult among the Menoninee of Wisconsin "pledge that the initiates will after death be guaranteed a good reception in the realm of the dead," and the Winnebago on Green Bay "believe that a whole series of reincarnations are secured by participating in the medicine society." In other cases, the gods are asked for help in this life. Among the Zuni, special societies may confine themselves for several days in a "cult room," dancing and singing to the gods to bring rain.[26]

Religion clearly provides an opportunity to appeal for supernatural benevolence as well as appeasing the gods to avoid their punishments—supernatural *rewards* are also common and important in traditional, ancient, and modern religions. The issue at stake, however, is whether punishment has special qualities that rewards cannot match. First of all, we should note an important asymmetry: The *denial* of rewards can itself become a punishment (for example, not living well enough to gain eternal life in heaven or a favorable reincarnation). Yet punishments cannot somehow become rewards. Therefore, when people look to the future, even rewards can become a deterrent if transgressions increase the chance that these rewards will be withdrawn (one can withdraw punishments as well, but that only *weakens* deterrence, rather than strengthening it). Even the carrot, therefore, can be used as a stick.

But as we saw in Chapter 2, there are several reasons why—in game theory, and in human psychology and behavior in general—rewards are weaker than punishments. And we can now explore whether this asymmetry plays out in the context of *supernatural* punishment. If supernatural punishments are more effective than supernatural rewards, it may be no coincidence that major theories of religion are overwhelmingly focused on the negative. Karl Marx saw religion as a tool of oppression. Edward Tylor and James Frazer saw it as a way of explaining death. Sigmund Freud saw religion as an obsession with a disciplinary father figure. Modern theories also seem to be built on negative elements. Evolutionary anthropologist Lee Cronk highlighted religion as a way for leaders to manipulate followers. Richard Dawkins sees religion as a parasitic meme. Even David Sloan Wilson's and Richard Sosis's theories on how religion boosts cooperation are argued to work because in-group members are able to avoid, exclude, or outcompete outsiders. One particularly interesting instance of this negativity dominance in theories of religion is Harvey Whitehouse's "modes theory." This focuses on the role of *negative* communal experiences in the origin of religious beliefs and practice. Why? Because it is particularly the traumatic experiences of painful and demanding rituals and rites of passage that most effectively bond groups together. Participants in indigenous rituals undergo fasting, isolation, beatings, torture, torment, scarification, and various other forms of hardship and deprivation (Whitehouse's "imagistic" mode of religiosity). The alternative, "doctrinal" mode is good at carrying religion far and wide in modern, literate, urbanized populations, but that is a relatively recent phenomenon—a cultural adaptation to the evolutionarily novel problem of large, anonymous societies. In the evolutionary *origins* of religion, it was the stick that did the work. Painful small-group rituals seem to have provided the more powerful social glue.[27]

There is a double whammy here. Not only do negative aspects of religion serve as important influences on behavior, but religion taps into fundamental aspects of human cognition that are particularly receptive and sensitive to negative input—the negativity bias outlined in Chapter 2. As Whitehouse stresses, negative events are not just important in and of themselves, but have dramatic effects on the human brain. Traumatic, negative experiences and emotions much more strongly encode vivid, enduring memories. Gods and religious practices may be nasty or nice, but we remember the nasty ones better. Since people are so sensitive to negative information and events, even if negative information and events are rare or not particularly emphasized by religious leaders, theologians, clergy, or peers, we are nevertheless especially disposed to detect them, worry about them, and react to them. Therefore, even if religious doctrine or practice emphasizes reward, human beings are likely to be more concerned with punishment.

The primacy of negative beliefs and practices in theories of religion, as well as in the empirical evidence from psychology, ethnography, cross-cultural studies, and laboratory studies, is just as we would expect if the human brain is primed toward negative phenomena, and if the negative consequences of supernatural punishment exert a special power. This negative bias may seem odd from a theological perspective, but makes sense from a psychological and economic one, because punishment has an intrinsic leverage—on human beings in general and on cooperation in particular. Rewards may encourage everyone to cooperate, but they cannot deter all of them from cheating. As we saw in Chapter 3, laboratory studies explicitly designed to compare the influence of supernatural reward and supernatural punishment on cooperation found negative consequences to wield a special power. As psychologists Azim Shariff and Mijke Rhemtulla put it, "though religion has been shown to have generally positive effects on normative 'prosocial' behavior, recent laboratory research suggests that these effects may be driven primarily by supernatural punishment."[28]

SYNTHESIS

So far in this chapter we have considered theoretical advantages of supernatural punishment in promoting cooperation, a diversity of beliefs about punishment in the afterlife and in this life, a variety of agents of punishment, and differing roles of supernatural punishment and reward. How can we make sense of all these variations and commonalities to draw out broader patterns from a complex web of beliefs?

VARIATION: PUNISHING MORE OR LESS

Supernatural punishment beliefs are remarkably diverse. The agents, punishments, timing, and targets can be remarkably different. Yet in most religious traditions it seems to pervade both life and death. One can find exceptions, of course. Supernatural punishment is not a concern for certain contemporary theologians, liberal Christian churches, or traditional societies. Such examples might be taken as evidence that the theory is wrong. Supernatural punishment is not important in *my* religion (says the believer), or in *my* people (says the anthropologist), or in *my* interpretation of the faith (says the theologian). But such cases are not a challenge to the theory, for two reasons. First, they appear to be exceptions to an otherwise general rule. As we have seen, ethnographic and cross-cultural studies suggest that beliefs in supernatural punishment are widespread, powerful, and deeply rooted, and this goes for indigenous, ancient, and modern societies alike. Exceptions do not reverse a broader trend.

Second, variation itself is an important part of the theory. Most aspects of human biology and behavior vary, just like everything else in nature. Darwin's theory of evolution was as much about variation in characteristics, selection among those variants, and the generation of new variants, as it was about any universal features of life. My theory is that beliefs in supernatural punishment were favored by selection because they helped to avoid the costs of selfish behavior and promote cooperation. If so, then in places or times when cooperation was difficult or in demand, we may expect supernatural punishment to come to the fore. And in places or times when the fruits of cooperation could be harvested just as well or better by other means, supernatural punishment may be superfluous or even counterproductive. An adaptive theory would therefore *expect* it to be reduced or absent in some circumstances, and accentuated in others (whether the adaptation arises from biological or cultural selection). In short, if belief in supernatural punishment has evolutionary causes and consequences, then it should vary depending on our prevailing social and physical "ecology"—just as other functional biological and cultural traits do. Richard Sosis and colleagues found harsher rituals among societies exposed to intergroup warfare, John Snarey found more moralizing gods among societies that suffer water scarcity, and David Sloan Wilson found specific religious beliefs helped to solve local problems of cooperation in a variety of historical groups. Likewise, supernatural punishment should occur more in some environments and less in others—more when cooperation is crucial, less when cooperation is not. Variations in supernatural punishment, therefore, are not falsifications of the supernatural punishment theory. They are predictions of it.[29]

If supernatural punishment is about reducing selfishness and promoting good behavior and cooperation, then a remaining puzzle is why there are some supernatural agents who do not merely fail to punish, but punish the wrong people. Why, for example, were some of the Greek gods so jealous, vengeful, and vindictive? Why in the book of Job did a perfectly good God exact apparently unjustified and arbitrary punishment on an innocent victim? Why are there even supernatural agents that work in opposition to each other? God and Satan is one obvious example, but the phenomenon is found elsewhere. The Greeks could appeal to one god for protection against another.[30]

Such examples seem to cut against the grain. While people continue to honor and revere capricious gods, they face the prospect of irrational and vengeful behavior, and may feel they can do little to avert it. Capricious gods certainly don't easily conform to the supernatural punishment theory. However, amidst the great innovation and diversity of human cultures, we should expect variation, cultural drift, even paradox. Statistical outliers are a feature of most general theories. What matters is the overall trend. While capricious gods crop up they are far less pervasive than the numerous supernatural agents that reward and punish for some systematic *reason*. We might even be able to predict where they should occur: In evolution, variation in a trait is greater when selection pressures on that trait are lower. One hypothesis, therefore, is that capricious gods emerge among groups where cooperation is less vital or easier to achieve. Either way, capricious gods are no more of a problem to a theory of supernatural punishment than the presence of corrupt politicians are to a theory of democratic government. With enough selection, or enough elections, the things that matter shine through.

Despite different means and methods, religions across the world and across history share a concern for some kind of supernatural consequences for people's actions. It is remarkable that so many different cultures in such distant and different regions have independently come up with exactly the same notion—that supernatural agents monitor them and punish socially unacceptable behavior. However, if such beliefs arise as a common solution to a universal problem—namely, a system of deterrence necessary to promote cooperation—then it may not be so remarkable at all.[31]

Punishment is closely linked to behavior that is seen as right or wrong. Some scholars have argued that only recent, world religions are truly moral religions. But this is only tenable if one takes a narrow, western view of morality. Whether people are concerned with God (like Christians), gods (like the Egyptians of antiquity), ancestors (like the Burkina Fasoans), or spirits (like the ancient Hawaiians), supernatural consequences tend to result from behavior that is perceived as "good" or "bad" for the community as a whole (and also often for the individuals themselves). Brown's human universals included "moral sentiments" and distinctions between good and bad and right and wrong. And Rossano found among indigenous societies that "ancestors are thought to be especially concerned with upholding social order through right behavior, observance of tradition, and avoidance of taboo, so as to ensure the community's continued fecundity and the security and legitimacy of its offspring." Now, what is perceived as right and wrong obviously varies among societies, but there is little reason to think that our version classifies as "moral" and other ones do not. As long as we maintain a modicum of cultural relativity, the link between supernatural punishment and moral behavior seems blindingly obvious. Anthropologist Bronislaw Malinowski suggested as long ago as the 1930s that "from the study of past religions, primitive and developed, we shall gain the conviction . . . that every religion implies some reward of virtue and the punishment of sin." What counts as "virtue" and "sin" will be very different among societies, as

Malinowksi perfectly well understood, but valuing whether they are broken or not is universal. And their function appears to be the same. Morality is very much grounded in the logic of cooperation among individuals. Good deeds help others and the community. Bad deeds harm others and the community. Supernatural punishment is typically about deterring selfishness and promoting cooperation, maintaining social order by fostering good relationships and reducing or defusing conflicts of interest.[32]

Morality is, then, not the preserve of western or modern societies. As political scientist David Welch put it, "cultures are extremely diverse, but the diversity of beliefs and practices among various cultures has proven to be somewhat misleading, because it masks much that cultures have in common. Comparative ethicists have shown that there are no premoral societies; that all societies give some degree of moral value to such things as human life, sexual restraint, friendship, mutual aid, fairness, truthfulness, and generosity; and that all societies employ moral concepts such as good, bad, right, wrong, just, and unjust." New work in experimental psychology corroborates the idea that all (mentally healthy) humans show the same fundamental grasp of and similar responses to ethical dilemmas and moral behavior. While there is variation in how much interest a given society's gods or other supernatural agents take in people's moral behavior, the overall trend is that moral behavior usually *does* tend to be linked to religious beliefs, and that is no coincidence. Morals can be promoted and enforced by other *human* members of the community, but evidently they can also—and often more powerfully and more effectively—be promoted and enforced by *superhuman* gods and spirits.[33]

CONCLUSIONS

If punishment is carried out by God, or other supernatural agents such as lesser gods, ancestors, or spirits, then many of the thorny game-theoretical problems of achieving cooperation are solved. Supernatural punishment is an extremely efficient deterrent because it allows constant and pervasive monitoring, severe and unlimited punishment, and shifts the burden of administering sanctions from people to supernatural agents. It also makes humans more equal. A strong and dominant individual may be somewhat immune to the reprimand of his tribe, but he is not immune to the reprimand of a god. Supernatural agents are the great levelers of human society. As far back as we can see into human history, and across the world's societies, the fear of gods or other supernatural forces has proven to be a compelling solution to the problem of human cooperation.

To achieve this, however, certain things must be in place: a set of rules that people are supposed to follow, a belief in supernatural agents or agency, and a belief that these supernatural forces reward and punish our actions. These are common—virtually universal—features of religions the world over and throughout history, and they seem to recur because they are what make religions work. Only supernatural agents with the power to affect our lives (and afterlives) are worth worrying about. As philosopher Aku Visala put it to me, "you can have a God without causal powers, but from a religious point of view he would be kind of useless." The common patterns of religion are no accident. It is the way the available building blocks must be approximately lined up if they are to be salient to human brains, and if they are to solve fundamental problems of human social life. People's widespread expectation of supernatural consequences for their actions may, therefore, be a quintessential ingredient of society—in the past, today, and in the future. As Voltaire suggested, "if God did not exist, it would be necessary to invent him." Human societies have invented gods not just once, but thousands of times.[34]

Having emphasized the doom and gloom of supernatural punishment, it is important to reiterate that the point is not to punish for punishment's sake, but to make people do something. And that something, we have learned, is typically cooperation. In turn, promoting cooperation is not an end in itself. Cooperation is sought because it allows us to achieve significant mutual benefits that we could never obtain on our own. So rewards are the ultimate end point even if punishment is the means to this end.

Later, we will examine exactly how a belief in supernatural punishment evolved. How could a belief that curtails self-interested behavior survive the ruthless mill of natural selection? But first we take a step back from religion, because it turns out that an expectation of supernatural consequences of our actions is not limited to believers, but extends to all humans—religious or not. Whether you are a believer, an agnostic, or an atheist, this book is about *you*.

CHAPTER 5

THE PROBLEM
OF ATHEISTS

Suspicion always haunts the guilty mind;
the thief doth fear each bush an officer.

—*Shakespeare*[1]

Cricket is a sport puzzling to many rational human beings, even in England where it was invented. It involves a litany of bizarre rules and vocabulary, requires both teams to dress in identical all-white outfits, and can take a whole day or more to play a game, with several breaks for lunch and afternoon tea, and additional ones for rain. Its popularity is even more puzzling when one considers that England has reliably been beaten by many other countries that have taken it up since. In 1882, the *Sporting Times* published a tongue-in-cheek obituary of English cricket after Australia beat England at the hallowed grounds of the Oval in London for the first time. The obituary said that English cricket had "died at the Oval on 29th August 1882" and that "the body will be cremated and the ashes taken to Australia." Ever since, the biennial competition between the two countries has come to be known as the quest to recover "the Ashes."

In the 2005 Ashes, England started to do pretty well after having lost repeatedly for the best part of two decades. The whole country became transfixed. People in my generation had grown up knowing that England regularly *competed* for the Ashes. But we didn't realize we could actually win. Might we really be on the way to victory? It didn't seem likely, but despite the long odds—or perhaps because of them—the British populace started to do everything they could to help swing the tides of fate. Bear in mind that the Ashes is a series of five "tests" or matches over the summer, each test lasting about five days, so there is plenty of time for anxiety and anguish. After five days glued to the TV, impervious to the glorious summer sunshine outside, a test can even end in a draw. For many non-cricket lovers, it is somewhere between hell and a test of one's sanity, or at least of one's marriage. By the end of the summer of 2005, however, somehow England won. Amidst the euphoria, people scrutinized how on Earth the feat could have been achieved. And people came up with some extraordinary answers. *The Times* posted a webpage asking the public "How did you help England win the Ashes?" The response was immediate and heartfelt.[2]

As Peter Thomas of Exeter reported: "On the final day of play I had to sit cross-legged on my bed, the right under the left. If I changed them, a wicket would fall [a 'wicket' means a batsman being caught or bowled out]. Needless to say I had two very dead legs." Andrew Smith from London similarly noted a link between his behavior and England's performance: "When I sat on the chair at my desk in my office, a wicket would fall. When I sat on the sofa, no wickets would fall and England would score runs. So, I moved the telephone over to the sofa, but I could not move the computer over, and whenever I went to answer e-mails, a wicket would fall. I did very little work on Monday." Andrew Sullivan from West Keal had to sit on the sofa doing a crossword. "Normally I finish the crossword fairly quickly," he said, "but if I pretended I couldn't solve a clue, then England kept the upper hand." Most of the responses were from men, but not all. Sarah North of Hertfordshire said, "I had become so nervous by the final morning of the series, that I couldn't sit still and watch the Test unfold. To make matters worse, every time I entered the room, Shane Warne would become the bowling God that he is and take an English wicket. By lunch time, I was banned from the house and had to listen to the remainder of the Test on the radio in my car, whilst driving in never-ending circles. I lost count of how many times I drove around the same route."

Paul Lochart from the Wirral felt the need to apologize to the team for failing to meet his ritual obligations: "Unfortunately I must travel through one of the Mersey tunnels to and from work. As it is impossible to listen to the radio whilst journeying through them, I am forced to break my otherwise uninterrupted listening. Each time I emerged

from the other side I found a wicket had gone down ... Sorry guys."
Fortunately, he was able to atone for his errors by finding an alternative way to exert ritual control. "To make up for this I went straight home, put on my England cricket shirt and sat in the exact same position on my sofa concentrating 100 percent on each and every ball, counting down overs and ticking off mini-targets," he said. "After it was all over I had to go straight to bed with the worst tension headache I have ever experienced, but all worth it and deserved penance for my careless journey home at so crucial a time." Numerous other responses flooded in, and everyone had a similar story to tell. So miraculous was England's victory, it seemed, that only magic could explain it.

THE VARIETIES OF SUPERSTITIOUS EXPERIENCE

English cricket fans may not be good representatives of *Homo sapiens* as a whole. But they are hardly alone in displaying superstitious behavior. Many of us, even if we declare ourselves atheists, nevertheless display a range of bizarre rituals and beliefs about the causes of events that have no material basis whatsoever. We'll explore lots of examples in this chapter. What strikes me as interesting is that these beliefs and behaviors are not random. Rather, despite their great diversity of content and context, they all tend to focus on attaining or avoiding some *consequence* or other. There is an assumed deep linkage of cause and effect in the way the universe works. Whether believers, agnostics, or atheists, we often seem to expect some kind of reward or punishment for our behavior, even when it is not materially possible. That is, we expect *supernatural* consequences. Do something the right way, and all will be well. Do something the wrong way, and we tempt disaster.

Psychologist Bruce Hood has conducted many years of research into this phenomenon and provides a litany of examples, not just from a few quirky individuals he has unearthed in laboratory experiments, but among the public in general and many well-known people. Tony Blair always wore the same shoes to the weekly chore of prime minister's questions in the House of Commons. As well as the lucky poker chip he carried in his 2008 presidential campaign, Barack Obama "developed a bizarre superstitious ritual of playing basketball on the morning of every election in his path to the White House." John McCain, Obama's rival in 2008, had a "catalogue of superstitions." For example, he carried a lucky feather and a lucky compass from the Vietnam war (although as Hood notes, "one wonders why, seeing as he was shot down and spent years as a prisoner of war"). He also carried a lucky penny, a lucky nickel, and a lucky quarter. The phenomenon extends to other walks of life and a range of cultures and upbringings—it's not just western men. British Comedian Shazia Mirza, who grew up in a Pakistani and Muslim community in Birmingham, in the United Kingdom, admitted "I've worn the same clothes at every gig for seven years. I worry that if I change my shirt it will all go wrong and no one will laugh." Nigerian born writer Chimamanda Ngozi was raised in both Igbo and Roman Catholic traditions, and though she finds superstitions "silly," she nevertheless admits that "sometimes when I trip, or nearly slip on the stairs, or painfully bump against furniture, I wonder if it means that God is getting back at me for something, anything; and if the near-accident happens in the morning, I think—fleetingly—that it might mean the day ahead will not go well." The note that it is "fleeting" captures the point that these thoughts cross our minds immediately and automatically, and they are only thereafter suppressed by our more logical, rational selves. Even in urgent situations, when one might think that people's attention would be directed elsewhere, we remember the dangers of breaking rules. When singer V. V. Brown found herself locked out of her house she "was still careful not to go underneath the ladder I used to get back in." Even psychologists, with the lowest rates of religious belief of any academic discipline, are not immune. Stuart Wilson noted in his own article on the psychology of religion that during one bad week his washing machine broke, his carpet got ripped, he caught a nasty cold, his kitchen counter fell off, and finally his toilet broke: "In exasperation, I raised both arms to the sky and screamed 'Why me?' It seemed like the natural thing to do, although I quickly regained my composure and became fascinated by

my response." Such patterns of thought are all-too familiar, historically persistent, and cross-cultural.[3]

A SPORTING CHANCE

Superstition is not a random collection of beliefs and expectations.
Instead, people are concerned with specific rituals or talismans that serve
to encourage positive outcomes and avoid negative ones. The phenomenon is perhaps especially common in sports. People do not necessarily
expect these actions or objects to be decisive, but they are an important
element of the activity surrounding competition. And if anything, athletes are especially concerned with avoiding the *negative* consequences
of failing to follow their ritual prescriptions. They can become quite
agitated if something spoils their rituals or if they lose a lucky charm.[4]

Baseball is rife with superstition. Turk Wendell of the Chicago
Cubs was voted the most superstitious athlete of all time by *Men's
Fitness* magazine: "He doesn't wear socks on the field. He waves at
the center-fielder before each inning. He brushes his teeth between
innings. He makes three crosses with a finger in the mound dirt before
he pitches. When the inning is done, he sprints from the mound, leaps
sideways over the foul line and spits out what appears to be four
pounds of black licorice. And he eats the same dinner at the same
restaurant chain the night before every start: French onion soup.
Peel-and-eat shrimp. Broccoli bites. Salad. Garlic sticks. Four-cheese
lasagna. And something called 'Death by Chocolate' for dessert." He
also wore a necklace with the teeth of animals he had shot, and when
he moved to the New York Mets in 1997, he asked for his contract to
be $9,999,999.99, in keeping with his shirt number, 99. Though an
extreme example, such behavior is commonplace in the game.[5]

One study of American and Japanese major league baseball players
found that 74 percent of professional players in both countries exhibited superstitious behaviors, and 53 percent of them did so in every
single game. Examples included "using a lucky toothbrush before each
game, eating chicken before each game, using the same shower before

each game, chewing three pieces of gum at the start of the game, drawing four lines in the dirt before getting in the batter's box, retying shoes during the sixth inning, taking hat off with the right hand only, and leaving glove in the same lucky spot on the bench. Wearing the same articles of clothing game after game (or until a bad game) was quite common." Remarkably, many continued these behaviors even though they were skeptical about their effectiveness. Only 27 percent of those who engaged in such superstitions said that it "always" or even "often" has an impact.[6]

Superstitious behavior is found in many other sports too. Michael Jordan wore his University of North Carolina shorts under his NBA shorts for his entire career. Tiger Woods always wears red on the last day of a tournament. Ernie Els will never use a ball with the number 2 on it. A variety of superstitions are found in tennis, affecting players such as Goran Ivanisevic, Andy Murray, Rafael Nadal, and Serena Williams.[7]

Bruce Hood's study found Serbian-born tennis player Jelena Dokic to be "the most complicated in her rituals, or at least the most honest and open about them." Any one of them would be odd in isolation, but in combination they add up to a remarkable web of obligations: "First she avoided standing on the white lines on court. [John McEnroe and Martina Hingis did this too]. She preferred to sit to the left of the umpire. Before her first serve she bounced the ball five times, and before her second serve she bounced it twice. While waiting for serves, she would blow on her right hand. The ball boys and girls always had to pass the ball to her with an underarm throw. Dokic made sure she never read the drawsheet more than one round at a time. Finally—and bear this in mind sports memorabilia collectors—she always wore the same clothes throughout a tournament."[8]

One might think all this could be perfectly rational behavior to help players concentrate or control their nerves—so-called preperformance routines in the sport literature. But much of it is openly superstitious. For example, Dokic herself says that the ball boys' underarm throw "is luckier than an overarm throw." Sports psychologist Aidan Moran noted, after considering the case of Dokic, that "clearly, this example highlights the fuzzy boundaries between preperformance routines and superstitious rituals in the minds of some athletes."[9]

Sports psychologists explicitly differentiate between routines and superstition. Routines allow one to establish control over the situation, whereas superstitions reflect the idea that fate lies *outside* one's control. And whereas routines may have a rational basis, such as improving concentration, superstitions are not *objectively justifiable*. Even routines that seem to verge on obsessive-compulsive disorders can sometimes be advantageous, because they create an illusion of control that can at least be psychologically comforting. But they can have pernicious effects. While routines can be shortened or omitted, superstitions "tend to *grow longer* over time as performers 'chain together' more and more illogical links between behavior and outcome." As Moran concludes, "the pre-shot routines of many athletes are often invested with magical thinking and superstitious qualities."[10]

What is especially remarkable—as we saw with cricket—is how even *viewers* of sports seem to fall prey to the same kind of superstition. It may be odd to wear your lucky socks on the court, but at least these athletes are actually involved and might derive some benefit to their performance or to their confidence. But hundreds or thousands of miles away, at the other end of a TV, we couch potatoes still do all sorts of strange things to bring on victory and avoid defeat. At the least, we all (right?) find ourselves gesticulating, shouting, and berating the players or the referee. But sometimes even fans lapse into much more specific obligations. In baseball one finds the phenomenon of the "rally cap," where whole sections of the crowd and any players on the bench turn their caps inside out when the team needs a boost. Notably, it becomes especially important when the stakes are high. Wearing your cap the usual way around could be disastrous.

PUBLIC DISORDER

But what about superstitions among the wider population? A 2005 Gallup poll in the United Kingdom found large proportions of people believe in supernatural events such as haunted houses (40 percent), astrology (24 percent), extraterrestrial aliens that have visited Earth (19 percent), communication with the dead (27 percent), and witches (13 percent). Another British survey in 2003 found that 74 percent of the public touch wood, 65 percent cross their fingers, 50 percent avoiding walking under ladders, 39 percent fear smashing mirrors, 28 percent carry a lucky charm, and 26 percent avoid the number 13 (and indeed many buildings have no thirteenth floor). I once went on vacation to Malaysia with some friends and we decided to save some cash by flying with Romanian Airways. That choice seemed fine at the time—we were 18-year-old students, who cared by what means we would be whisked to our tropical beach paradise? When we came to change planes in Bucharest, however, my friend Dennis and I noticed we had seat numbers 13A and 13B. O.K., perhaps now we were tempting fate a little. But before we could dwell on what might be in store for us on our flight, we discovered that the Romanians had thoughtfully removed the entire thirteenth row! Problem solved. Well, except for the subsequent chaos to squeeze everyone in.[11]

Supernatural beliefs also run high in America. A US Gallup poll in 2005 found that 41 percent of people believe in extrasensory perception, 37 percent believe in haunted houses, 32 percent in ghosts, 31 percent in telepathy, 26 percent in clairvoyance, 25 percent in astrology, 21 percent in communication with the dead, 21 percent in witches, 20 percent in reincarnation, and 9 percent in spiritual possession. Fully 73 percent of people believed in at least one of these supernatural phenomena.[12]

And the high rates of belief in supernatural phenomena are not just quirks of the times. These figures have not changed much over many years of similar surveys. Numerous other aspects of life show us that superstitions are big business. Newspapers the world over take up valuable real estate on their pages to print horoscopes. Palm readers, fortune tellers, Tarot card readers, and other supernatural services are present in more or less every city. Novels and guides on the paranormal fly out of bookshops. Superstitious behaviors crop up in all sorts of other personal and commercial activity as well, from selling houses (where it is common to bury a statue of St. Joseph upside down in the yard) to games of chance (where it is common, for example, for people to kiss dice before throwing them).

People often don't like to admit to being superstitious and can be particularly averse to the idea that they have supernatural beliefs, with the often accompanying stigma of ignorance, gullibility, or lack of education. But the beliefs and behaviors are out there for all to see. As Hood writes, "so long as no one mentions the word 'supernatural,' adults are quite happy to entertain notions of hidden patterns, forces, and essences." However, it does mean that people are likely to deny or downplay supernatural beliefs when responding to survey questions in polls like those mentioned earlier. The real rates are likely to be even higher. A more accurate way to explore supernatural beliefs may therefore be laboratory experiments, where expectations about causes and effects can be examined without having to rely on people reporting or even recognizing their beliefs themselves.[13]

Laboratory experiments find considerable evidence for superstitious and supernatural thinking, even among self-declared atheists. For example, Princeton psychologist Emily Pronin and colleagues conducted a series of manipulation experiments cleverly revealing that regular people tended to believe that they had influence over events even when that was impossible—believing, for example, that they had helped someone else net a basketball simply by willing the result, or that they had harmed someone by sticking pins in a voodoo doll. A raft of similar experimental evidence offers important demonstrations that while many of us may be atheists, few of us are immune to supernatural thinking.[14]

Finally, it is worth noting that *religious* beliefs, as well as mere superstition, may be much more common than appears at first glance. This is because they are held even among people who explicitly claim no allegiance to any particular religion. Pew Forum surveys have shown that a mere 5 percent of Americans "do not believe in God or a universal spirit," and even within this tiny minority only 24 percent self-identified as atheists (the rest presumably believe in something else, or remain unsure). A Baylor University survey found that 63 percent of Americans who have no religious affiliation nevertheless "believe in God or some higher power," and 48 percent admit to belief in an afterlife. Another study found that 34 percent of *atheists* expected to be called before God on Judgment Day. Supernatural beliefs may be especially obvious in religion, but they are widespread among the unaffiliated and even among atheists too.[15]

NO ATHEISTS IN FOXHOLES

Superstitions are rampant in everyday life. But they can become especially strong in times of crisis or danger. At these times, even the most ardent atheist can succumb to doubt and superstition. Detailed research has shown that people are consistently more likely to display superstitions under four specific conditions: high stakes, uncertainty, lack of control, and stress or anxiety (Figure 5.1). In the context of evolution, it is precisely high stakes and stressful events that tend to have consequences for survival and reproduction, and therefore superstition

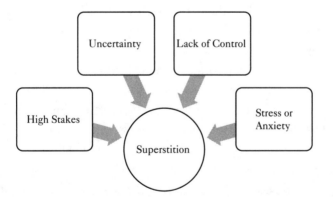

Figure 5.1 Superstition is common in everyday life, but is exacerbated in contexts of high stakes, uncertainty, lack of control, and stress or anxiety. A key element of superstition is the expectation of supernatural consequences of one's actions which, as shown in the cognitive science of religion literature, is a feature of human nature found in religious believers, agnostics, and atheists alike.

may have played a special role in influencing our behavior when it matters for Darwinian fitness.

The classic example in which these conditions come to the fore is war. As the saying goes, "there are no atheists in foxholes." This refers to the idea that a soldier cowering in his dugout, with bullets and shells flying all around him is, if ever, going to need to believe in something now. He needs all the help he can get, plus he'd better have his beliefs sorted out in case he is killed. It might even help him fight more confidently and effectively.[16]

But you do not have to go to war to find situations that amplify superstition. The examples of Tony Blair, Barack Obama, Tiger Woods, and Jelena Dokic may have something in common. The domains of politics and sport demand high performance under unrelenting scrutiny, and yet contain a large element of chance. In these professions, therefore, we also have the perfect storm of high stakes, uncertainty, lack of control, and stress or anxiety. We should therefore expect ritual and superstition to be especially prevalent. Even highly capable people come to appreciate the limits of their ability to control causes and effects, and when the fate of an important project is at least partly out of their hands, they may resort to unconventional means.

The core point that superstitions become prominent in times of angst can of course pertain to many events in normal life as well. Take travel. Many of us are familiar with St. Christopher, the patron saint of travelers. People often have an image or figurine of him in their cars. But the phenomenon is common across cultures. When writer Alana Newhouse began an extended book tour that involved twenty-five flights in six weeks, she found herself compulsively reciting the Wayfarer's Prayer (or *Tefilat HaDerech* in Hebrew) before take off. The prayer is supposed to protect travelers from danger. Thinking through the logic of her behavior, she remembered her high school study of the orthodox ArtScroll prayerbook: "The Sages teach that a person's sins are more likely to be recalled against him when he is travelling." With lots of time to reflect on this as she crisscrossed the United States, she reports, "I became obsessed with my sins, seeing as they were supposed to be in fuller view during travel. I can say with certainty that I have never been a better Samaritan than I was in the airports of America. I helped women maneuver their strollers, instructed the elderly on how to get through security and volunteered to take a middle seat so the two newlyweds could at least sit across the aisle from each other." And given an interesting feature of modern travel—unlike going independently by foot or donkey, these days we are trapped in same vehicle with numerous strangers—she also became suspicious of other people because *their* sins would become important to her own fate as well: "Did the man in 21A ever cheat on his wife? What about the Cruella De Vil-looking woman next to him? And the kid in 4D looked distinctly like the kind who'd have put a cat in the microwave. I did not want to go down for someone else's crime." After one especially turbulent flight, she found herself clasping the hand of her neighbor and they fell into nervous conversation. After landing, her fellow passenger asked, "why do I get so nervous? It's almost as if I think I deserve to be punished or something." Many people comment on the same thing. As anthropologist Scott Atran asked, "why do we cross our fingers during turbulence, even the most atheistic among us?" It's a feeling we all know: If something were to happen, why me, why now?[17]

My colleague Richard Sosis went to work in Israel and was surprised to find that the bus he took to his fieldsite would routinely pull over at the side of the road, specifically so that the driver could recite the Wayfarer's prayer over the intercom. At the end, the passengers would all say "Amen" (as is their obligation under Jewish law). With the prayer done, the bus would pull back onto the road. Sosis is an anthropologist who, with W. Penn Handwerker, was studying how people responded to the missile attacks launched by Hezbollah into Israel during the 2006 Lebanon War. We noted above that superstitions tend to become prominent when people lack control, and this is especially the case with unpredictable *punishment*. Stress is dramatically increased if we know that something might hurt us but we don't know when. It appears to be a deeply ingrained phenomenon. Experiments on both rats and humans find that initially, we develop an illusion of control in an attempt to exert some influence over the situation. But when no control is possible and unpredictable punishment continues, animals and humans fall into a state of "learned helplessness" in which they become greatly depressed and unresponsive—even missing opportunities to help themselves out of the situation.[18]

Sosis and Handwerker were particularly interested in how religious rituals might help people in dire circumstances feel some sense of control and cope with the stress. The recitation of pslams, among other rituals, was a good candidate because they "are believed to protect performers from dangers such as warfare, voyages, sorcery, and disease." The duo recounted a Jewish joke about a Hasid who, in the midst of an emergency, calls out: "Jews, don't just stand there and depend on miracles. Do something! Recite psalms." But the recitation of psalms is taken very seriously. As Sosis and Handwerker describe, prayer books often have a list inside the cover, specifying which psalms should be read for what situation, such as "traveling, depression, healing, birth of a child, livelihood, peace, finding a mate, and countless other situations in which one might wish to either beseech or thank God."[19]

What Sosis and Handwerker found was remarkable. Women who remained in the danger zone in Northern Israel—where a missile might fall out of the sky at any moment—showed a significant reduction in levels of anxiety if they recited psalms as part of their everyday rituals. A control group of women from the same area, but who had relocated out of the danger zones, showed no relationship between psalm recitation and anxiety. All 115 of these women were Orthodox Jews, setting them apart from the purely superstitious types of behaviors discussed earlier, but the point is that ritual served a purpose in dealing with uncertainty. And the rituals were begun or increased in direct response to a threat. The women would sometimes ensure that the entire set of 150 psalms in the book were read every day, or even three times a day, by assigning different sections to individuals in a group.

But do people really think it helps beyond just calming their nerves? In previous research Sosis has found that 78 percent of (religious) people "strongly agreed" (the highest point on a 10 point scale) that "reciting psalms can improve the situation [of the intifada]," and 69 percent strongly believed that "reciting psalms can protect one from an attack." Not only this, but reciting psalms was believed to have "much greater efficacy than government actions" in improving the situation. Although these studies were of religious Jews, Sosis notes that "the Book of Pslams has deep resonance even among secular Jews." In fact, he got the idea for the study from a secular Israeli cab driver who had a copy of the book of psalms on his dashboard "as protection while driving through dangerous areas."[20]

Similar effects have been found in other studies. During the 1991 Gulf War, Iraq fired a series of Scud missiles into Israel. One study compared people who lived in high-risk areas from those in low-risk areas, and while it was not the primary focus of the study, the researchers noted that people from high-risk areas knocked on wood during interviews more than those from low-risk areas. The implication was that uncertainty and risk had driven them to higher levels of superstition—presumably in a subconscious effort to exert control over an inherently uncontrollable situation.[21]

The psalms study seemed to suggest a perfect illustration of the eminent anthropologist Bronislaw Malinowski's theory of magic. Magic has many meanings in modern society, but it gained scholarly attention in the early part of the twentieth century when Malinowski began to examine magic from a scientific perspective. In his research among the Trobriand Islanders in the Pacific, he noticed that magical thinking and behaviors were dominant in activities surrounding open sea fishing expeditions, which were dangerous and subject to great variation in success or failure, but absent when men fished in lagoons, which were safer and provided lower but more stable returns. This basic relationship—a situation of uncertainty leading to or increasing superstitions—has since become known by anthropologists and psychologists as "the uncertainty hypothesis." It seems that magical rituals can be advantageous in coping with unpredictable situations. A variety of studies have found this effect among gamblers, students taking exams, golfers, baseball players, and victims of warfare. In all such cases, research suggests that establishing a sense of control can be effective at reducing stress and anxiety. It may be an old idea, but magical thinking is alive and well.[22]

ANIMAL ORIGINS OF SUPERSTITION?

Humans might seem to be unique, perhaps even defined, by our super-natural beliefs. But in fact there is evidence that supernatural beliefs, or at least beliefs that associate cause and effect even where none exists, have significantly older evolutionary roots.

Harvard psychologist B. F. Skinner achieved fame for his championing of behaviorism, the idea that all behavior can be elicited with the appropriate conditioning. His classic experiments were conducted in what became known as Skinner Boxes, in which he put a rat with a lever and a food tray. The rat would initially press the bar by accident, resulting in a food reward falling into the tray. Eventually, the rat would learn this association and keep pressing the lever until it was satiated. O.K., big deal. But things got interesting when he did a similar experiment with pigeons. They had to perform a similar action to obtain a food reward. However, Skinner found that even when the food was dispensed at random, the pigeons continued to repeat the ritualized behavior of pressing the lever as though they believed this was the cause of the food arriving. In effect, the pigeons seemed to be displaying superstitious behavior, just like English cricket fans.

Skinner's pigeon experiments came under a lot of criticizism—could animals really be "superstitious"?—but similar effects have since been found in other animals as well as in humans. What is especially interesting for our purposes is the idea that natural selection might have actually *favored* such behavior. Evolutionary biologists Kevin Foster and Hannah Kokko developed a mathematical model to explore the effects of superstitious behavior on biological fitness, and found that overestimating the association between the cause and effect of events can be adaptive. For example, because the costs of predation (being killed) far exceed the costs of hiding from a possible threat (a short period of inactivity), evolution should favor mechanisms that *tend to assume* that any noise signals a predator, even when the chance is remote. Although making such false associations may incur costs, it is much *more* costly to err in the other direction. As Foster and Kokko explained, "natural selection can favor strategies that lead to frequent errors in assessment as long as the occasional correct response carries a large fitness benefit."[23]

Of course, we have to be careful to distinguish between animal behavior and more complex human thinking. As Foster and Kokko noted, Skinner "focused on there being an incorrect response to a stimulus (behavioral outcome), rather than the conscious abstract representation of cause and effect (psychological relationship), with which human superstitions are often associated." Since biological fitness is a result of behavior, not thoughts, however, Foster and Kokko followed Skinner's lead and modeled behavioral outcomes without making any assumptions about psychological beliefs. After all, how can we know what animals really think? But in fact this approach makes their model more widely applicable: without any assumptions about cognition, the logic applies to any organism—whether rat, pigeon, or human—regardless of whatever psychological mechanism may be giving rise to it. The effect was powerful enough for them to conclude that superstitious (or superstitious-like) behaviors are "an inevitable feature of adaptive behavior in all organisms, including ourselves." Superstition, therefore, should not be regarded as surprising, or errant, or even human. Of course, if taken to extremes, superstitious behavior can become debilitating, as occurs with obsessive compulsive disorder, when people cannot stop thinking or acting in certain ways even if it is harming them. But some moderate level of superstition appears to be a universal aspect of human nature. Erring on the side of caution is an effective survival strategy, even if it leads to irrational beliefs and bizarre behavior.[24]

THE EMERGING PATTERN: SUPERNATURAL REWARD AND PUNISHMENT

So far we have seen how people in general harbor a variety of superstitious beliefs and behaviors, but all of which are aimed at one thing: managing supernatural reward and punishment. There is an assumption that *if* you do X, then Y will happen, or *if* you do not do X, then Y will not happen, or—perhaps worse—Z will happen instead.

(The distinction is important because punishment can be about being denied Y as well as suffering Z.)

In other words, although superstitions come in all shapes and sizes, they are all attempts to exert control over events. We want good things to happen and bad things not to happen, and we try and find ways to influence these outcomes with our behavior. Since we cannot always control them materially, we must turn to nonmaterial means instead. This corresponds to a belief system in which we anticipate—explicitly or implicitly, consciously or subconsciously—supernatural rewards and punishments for doing certain things and not others. We might deny that we *really* believe such actions have consequences, but across cultures and throughout history people have found themselves remarkably reluctant to tempt fate. As long ago as 1757, Scottish philosopher David Hume noted that "by a natural propensity, if not corrected by experience and reflection, [we] ascribe malice and good-will to everything that hurts or pleases us ... the inanimate parts of nature acquire sentiment and passion."[25]

The central role of supernatural punishment and reward in superstitious behavior can be seen in Bruce Hood's research, which focuses on why people associate objects with desirable or undesirable traits. He achieved some notoriety for giving public talks where he would offer people the chance to hold a pen that was purportedly Einstein's (everyone wanted to touch it), or to try on a cardigan that belonged to a renowned murderer (no one wanted to do it). Even though Hood was not looking at supernatural punishment per se, he nevertheless found a pattern in which lingering superstition focuses on consequences—and *negative* consequences in particular: "It can lurk away in the back of our minds whispering doubt and warning us to be careful. It can be that uncomfortable feeling we experience when we enter a room, or the conviction that we are being watched by unseen eyes when no one is there." People often appeal to superstitious beliefs and rituals to obtain or assure positive outcomes, but appealing to them to avoid *negative* outcomes appears to be an especially common and powerful concern. What we really want is to avoid Z or avoid being denied Y, and we do all we can to achieve it.[26]

Many people admit to such anticipation of punishment and reward and cannot fully rationalize it. We may recognize our rituals and their underlying beliefs as irrational and bizarre. Yet we often find it hard to sidestep them or give them up, especially when times are very good or very bad—since it is at precisely those times that we actively try to make sense of our luck. One striking confession comes from writer Alain de Botton, who declares himself an atheist but nevertheless admits to how the trials of life seem to connect tightly with a cosmic order of reward and punishment:

> I suffer from a basic superstition that you're allowed only so much good fortune before something very bad will happen—and, correspondingly, if things have gone wrong for a while, you'll be due an upswing in your fate soon. This makes me very wary of moments when I should, apparently, be celebrating. Holidays where the view is perfect and the weather ideal really worry me. Scenarios of disaster haunt me: something appalling is bound to happen soon. Professional success can be just as alarming. Storming up the bestseller list can be a reason to hide in bed in despair. I feel that, by being miserable before anyone tells me to be, I will escape the jealousy of higher forces. I am trying to disappoint myself before the world gets a chance to do it rudely for me. As for when things go wrong, I like to wallow and exaggerate the misfortune, for only when I'm at rock bottom can I have a sense of that now something or someone will smile more benevolently on me.
>
> Behind such absurd ideas, there's an even stranger faith in the interconnectedness of events. So I believe—without admitting to myself that I'm doing this—there is some connection between the disappointment I suffer at the hands of a publisher at 11 a.m. one morning and the piece of good news that comes the following evening. I feel as if I have "earned" the positive event in the eyes of something or someone who is keeping a giant ledger in the sky.
>
> Even more striking is that I am, on the surface, entirely committed to atheism, with a mocking scorn for those who would have a moment's patience for such things.[27]

Humans have a number of conscious and subconscious theories about how the world works, and our place within it. A significant chunk of them appear to revolve around an intrinsic expectation of supernatural agency and supernatural consequences. Hood calls this our "supersense" (lumping superstitious beliefs and beliefs in supernatural agents together as related phenomena): "But you don't have to be religious or spiritual to hold a supersense. For the nonreligious, it can be beliefs about paranormal abilities, psychic powers, telepathy, or any phenomena that defy natural laws . . . Even beliefs about plain old luck, fate, and destiny." It seems that a recurrent and powerful theory that people tend to hold onto is that there is an underlying cosmic order. Good begets good. Bad begets bad. Life is not a random series of events. It rewards and punishes. Despite being just one of 7 billion people, and in an infinite universe, we tend to think that whatever *we* do, as an individual, is noticed by some greater cosmic force, that the force cares what we do, and that it will keep track of us and tailor an individualized response. It's not unlike, of course, a belief in an all-powerful God.[28]

The key point here is that atheist superstitions come down to concerns for supernatural reward and especially supernatural punishment, just like religious beliefs. It is clearly something broader about human nature. But do we have any evidence that superstitions are a force for *good*? Or are they just arbitrary or even malicious? On the face of it, they can certainly seem largely self-interested—I do X or Y to improve *my* luck or fortune, whatever its effects on other people. In fact, however, while they vary enormously, on the whole superstitious beliefs tend to have a prosocial—even moral—dimension. When people want to obtain or avoid something important, they often appear to be particularly concerned to act in ways that are nonselfish and cooperative, largely because good behavior is seen to increase the likelihood of positive outcomes and mitigate the possibility of negative outcomes. We saw earlier how Alana Newhouse found herself being suddenly helpful and generous during her grueling schedule of flights. But the effect has been examined in controlled experiments as well. For example, one study found that when people were waiting for the results of an interview or a medical test, they were more likely to help others "as if they can encourage fate's favor by doing good deeds pro-actively." They were more willing to donate time and money or make generous pledges to charities. Benjamin Converse and his colleagues called this the "karmic-investment hypothesis," since people were act-ing as if they were building up a reservoir of good karma that might increase the probability of good news on their job or health. Not only did people act this way, but they also seemed to show a genuine belief in its efficacy. When they carried out such positive deeds, people's opti-mism about the desired outcome of their appointment increased.[29]

Superstitions are pervasive in life, especially when the stakes are high. But superstitions are not morally vacuous. Although we may employ them to fulfill our egoistic needs and desires, the way in which these goals are achieved is—not always but often—through supersti-tious behaviors that help others as well as ourselves. It is good deeds, not bad, that steer fate in our favor. Superstitious beliefs in supernatu-ral punishment, then, just like religious beliefs in supernatural punish-ment, can be a force for good.

WHAT ATHEISTS BELIEVE: THE COGNITIVE SCIENCE OF RELIGION

Ever since Plato we have liked to think of ourselves as rational beings, able to objectively study and learn the material nature of the universe, our society, and ourselves. The Enlightenment and the rise of science has only strengthened this view. But the idea of humans as rational beings is only an ideal. We are far from rational—not least, as Hood cautions, in the realm of supernatural thinking, where "common supernatural beliefs operate in everyday reasoning, no matter how rational and reasoned you think you are." If rational choice theory did anything useful, it was to provide a baseline against which a new wave of science could scoff. Legions of experimental psychologists and behavioral economists have shown that people deviate markedly and reliably from this rational ideal. Humans are inherently biased in numerous ways—in cognition, motivation, emotion, information processing, decision-making, learning, and memory. Superstition is just one notable example in a sea of irrationality.[30]

But psychological biases are not random, and researchers have begun to identify strong patterns in the kinds of things people tend to believe or not believe. You can take your pick of topic—how politicians decide whether or not to go to war, how doctors decide what treatments to prescribe, how people decide to invest their money—and you will find that first, the rational choice model is severely challenged, and second, that psychology can make some good predictions for how these people will actually behave in real life. Religion is no exception, and a large tranche of new research in experimental psychology suggests that we should not be surprised at all about the pervasiveness of supernatural beliefs. A whole new field called the "cognitive science of religion" has grown up around the observation that human brains seem predisposed to supernatural and religious thinking. It may not be rational, but it is, empirically, part of our biology. Academics always find details to argue about, but the evidence for a set of general cognitive mechanisms that underlie supernatural beliefs is overwhelming.[31]

This research is especially important because it looks at fundamental cognitive features that transcend different cultures and religions, and thus appear to be deeply rooted universals of human nature. It is crucial to recognize these basic cognitive underpinnings, because "while all religions come from culture, this is not true for all supernatural beliefs." Supernatural beliefs are deeper, older, primordial. They form the cognitive foundations on which religion not only stands, but relies. So what are they? And what are the implications for supernatural punishment?[32]

CAUSE AND EFFECT REASONING

One of our most fundamental attributes as human beings is a tendency to associate cause and effect: sometimes correctly, sometimes not. But we have a strong disposition to draw links nevertheless. This makes us highly susceptible to believing that one thing causes another even if the evidence is tenuous and the data are few. Once again, there is an asymmetry here in that our cognitive mechanisms appear to be designed not simply to attain accuracy, but to make sure we avoid *negative* events. As we saw in Foster and Kokko's study, this can make us highly sensitive to dangers in the environment. Michael Shermer stresses that "we have evolved brains that pay attention to anecdotes because false positives (believing there is a connection between A and B when there is not) are usually harmless, whereas false negatives (believing there is no connection between A and B when there is) may take you out of the gene pool. Our brains are belief engines that employ association learning to seek and find patterns. Superstition and belief in magic are millions of years old, whereas science, with its methods of controlling for intervening variables to circumvent false positives, is only a few hundred years old."[33]

One consequence is that we are disproportionately impressed by coincidences—strange events that seem beyond explanation unless they were somehow meant to be. Humans are not good at dealing with probabilities, so we find ourselves surprised by events that seem unlikely but are in fact mundane in a world of large social networks and millions of people. There are so many possible events and interactions in our own lives, let alone among all of our friends and acquaintances, that coincidences are everywhere, and they stick with us precisely because they often seem so bizarre.

Shared birthdays and knowing the same people are common examples. But here's a more extreme one. Two friends of mine have been happily married for some years. One day they were looking at a friend's photo album and came across some photos of a long-forgotten Oxford ball they had both attended as students, but before they had ever met. Looking closer, they were stunned to realize that in one of the photos, they were standing right next to each other, looking in opposite directions! They did not actually meet for another two years. How weird is that? The answer, even in this case, is: not very.

Critical for our purposes is the link between cause and effect reasoning and our "theory of mind." Theory of mind is the capacity to understand that other people know or don't know things—and that they may have knowledge similar or different to ours. This was a vital skill for navigating the increasing social complexity of our ancestral environment, as we will discover in the next chapter. But an additional element of theory of mind is that beyond mere knowledge, it comes packaged with an "intentionality system"—we can also surmise the *intentions* behind people's actions. This is important, because it means that not only do we search for links between causes and effects, but we also have an innate tendency to attribute positive and negative life events as happening *for a reason*—for example, "he did that *because* of what I did". And of course this applies not only to the minds of human beings, but also to the minds of gods and spirits. As soon as we begin attributing events to supernatural agents, we start to wonder why they may have caused or allowed them to occur—did this happen *because* I was bad, or good?[34]

The attribution of cause and effect is rooted in the way human minds work, and theory of mind helps us pin the blame on specific intentional agents. But there are cultural factors as well, which may be important for understanding cultural diversity in religious concepts. Richard Nisbett argues that the whole western approach to perceiving the world stems from a commitment to a cause-effect rationale that began with the Greeks, became solidified in western philosophy, and now pervades life in western capitalist societies. This worldview places a heavy emphasis on personal agency, so people get used to believing they have the power to control their own lives. Westerners tend to look for and identify agency. This is in contrast to eastern modes of thinking, with roots in Confucius, Chinese philosophy, and collectivist culture, in which people are less concerned about personal agency and more concerned about exhibiting self-control in order to fit in with the broader community of which they are a part. In such cultures, people get used to believing that their individual power to control the world is limited, and easterners often tend to assign agency to levels beyond the individual. This fits with the eastern focus on community obligation and karmic forces rather than a single God.[35]

While there is interesting variation, our underlying cause and effect reasoning appears to have deep evolutionary roots and derives from universal cognitive mechanisms. Ultimately, it offers an explanation for why people tend to assume that even random events in life have causes—whether they are attributed to a supernatural *agent* like God, a supernatural *agency* like karma, or superstitious concepts like fate.

MIND-BODY DUALISM

French philosopher Rene Descartes argued that the mind and the body are fundamentally separate. Even though the brain is a physical thing that can be examined, the mind, by contrast, is something we cannot observe and yet is responsible for complex phenomena such as consciousness and self-awareness. The fields of psychology and neuroscience have for several decades been chipping away at Descartes's model, finding ever more intricate links between, on the one hand, our thoughts and feelings, and on the other, tangible physiological structures and processes in the brain. At the extreme, some research suggests that even free will is an illusion, because material processes in our brain predetermine our behavior before we really "decide" for ourselves what it is we will do. The human mind may be just a representational extension of the brain, albeit a remarkably sophisticated one.[36]

Philosophers and scientists will no doubt continue this debate for years to come. But for our purposes, it doesn't matter whether Descartes was right. What matters is whether we act *as if* he were. And the results here are consistent and strong: all of us show an innate mind-body dualism—a tendency to believe that the mind and the body are fundamentally different kinds of entities. Moreover, this tendency is not something we learn from our cultural or religious upbringing. On the contrary, the belief appears to be in place from very early on in childhood; hence the title of developmental psychologist Paul Bloom's book, *Descartes' Baby*.[37]

The fact that we view things this way has a number of important implications. First, it makes it easier to believe in supernatural agents being "out there" because we can conceive of minds without having to observe any physical body that houses them. Without this basic cognitive capacity, the human conceptualization of God (and many other supernatural agents) would be dead in the water. Second, mind-body dualism allows different beliefs about life and death. The body eats, sleeps, works, plays, and dies. The mind thinks, wonders, knows, but does not die. Experiments have found that both adults and children show a natural understanding of the idea that the body ceases to function after death, but a remarkable inability to conceive of the same thing happening to the mind. Jesse Bering and David Bjorklund used a puppet show to examine children's intuitive beliefs, in which a mouse was gobbled up by an alligator. They found that even though children believed that the dead mouse no longer needed to drink, eat, or sleep, he still had thoughts and feelings. This cognitive stance emerges early on in development, and even adults display some of the same subconscious unwillingness to believe that death has erased the mind. For example, in Bering and Bjorklund's experiment, the belief that the dead mouse continued to think was the same among children and adults.[38]

Inasmuch as it boosts belief in supernatural agents and life after death, mind-body dualism contributes to the salience of supernatural consequences of our actions.

THE HYPERACTIVE AGENCY DETECTOR DEVICE (HADD)

The director of an institute I once worked at joked that he had a daily "barometer" of his performance. When he entered the office in the morning, he would pass a large, formal portrait of the founder of the institution. Sometimes the founder appeared to have a disapproving scowl, but other times he appeared to have a faint glimmer of a smile, suggesting a nod of approval. The picture, of course, did not change, but our perceptions can easily lead us to believe that supernatural forces are out there, observing and judging us.

Humans have an innate disposition to perceive agency in the environment, whether it is there or not. This is our so-called hyperactive agency detector device (or HADD). As ridiculous as it sounds, or perhaps precisely because it sounds ridiculous, the HADD has become a common topic of discussion among scientists, religious studies scholars, and theologians alike. The phenomenon has been carefully demonstrated in psychological lab experiments. In one study, researchers created a computer program that showed two little triangles moving around the screen in one of three patterns: randomly, as if they were trying to do something, or as if they were trying to deceive each other. What was interesting is that people tended to attribute the triangles' movements to intentional behavior—such as chasing, fighting, or tricking each other—even when they moved around the screen at random. Adults did this as well as children, ascribing agency even to chance events. The HADD appears to be an adaptive bias because in our ancestral past the costs of missing a predator, a malicious human, or some other danger are great. The HADD puts us on high alert for any kind of malevolent agent that could be detrimental to fitness. This obsessive social radar may be especially important because being taken by surprise can be a great leveler of power differences. You may be the strongest person in the tribe, with the strongest allies, but that may not help you survive an ambush by weaker but more cunning opponents or predators. Being on constant high alert for agency in our environment is not ideal, because it means that we will sometimes spend time averting danger where none exist, but this is a small price to pay to make sure we avoid the deadly risk of assuming there is no threat when there is.[39]

Agency detection has important consequences for behavior. Experiments show that priming people with the possibility that invisible agents are present can decrease self-interest and increase cooperation. For example, one study asked volunteers to complete a set of mathematical problems on computers. They were told, however, that there was a glitch in the computer program that sometimes accidentally showed the correct answer on the screen. If this happened, they were instructed to quickly hit the spacebar to avoid seeing the answer, so as not to mess up the purported experiment. The interesting part is that some of them were told, in passing, that the experiment had been designed by someone who had recently died, and whose ghost had been reported in the building. The results were remarkable: Individuals who had been primed by this story were significantly faster in hitting the spacebar to hide the answer, compared to participants in a control group. People were more reluctant to cheat in the supposed presence of a ghost.[40]

We appear to be particularly sensitive to the presence of eyes, which are thought to subconsciously activate concerns for reputation. In a now famous series of experiments, it has been found that people are much more cooperative when there are images of eyes in the room. One experiment deliberately included a very un-humanlike robot, except with very humanlike eyes, and found that people contributed significantly more in a public goods game when the robot was present. Another tested these findings in the real world. Melissa Bateson and her colleagues at Newcastle University ran an experiment in a university coffee room that had an honesty box for payment. They found that in weeks where they stuck a picture of eyes on the box people were much more likely to pay up compared to weeks with a neutral picture. The Bateson study has been criticized on a variety of grounds, somewhat unsurprisingly given the informal setting and various uncontrolled variables. But the basic pattern that the presence of eyes reduces selfish behavior and increases cooperation has been found repeatedly in a range of controlled laboratory experiments. It also appears to be a strong effect. Concerned about whether people might simply be responding to the presence of another *person*, or some other social factor, other experiments showed that cooperation increased even when the stimulus was mere spots arranged on a wall, with no obvious human face at all. Three spots arranged as two eyes and a mouth has been shown to activate the fusiform face area (FFA) of the brain, a hard-wired mechanism that allows us to recognize faces, which babies display within minutes of birth. Such experiments raise many questions, but should we really be surprised by the results? It is perfectly intuitive that when we are watched we behave differently.[41]

Out in the real world, we see a range of examples of this phenomenon. People behave in one way when they believe they are in private, and in another way when they are in public. And we are uncomfortable when norms of privacy are broken. A few years ago a novel kind of public toilet, with one-way glass, was placed in the middle of the street in a Swiss city. No one could see in from the outside but, from the inside, it appeared as if one was in a crystal clear glass box with cars and people whizzing by just inches away. Not surprisingly, the contraption caused endless comment and fascination and people found it extremely difficult to convince themselves that they were in privacy. It didn't matter whether people *could* see you or not as much as the *perception* that they could.

We also see a great concern for the presence of eyes historically. Various traditions, folklore, and religious concepts revolve around the power of eyes and observation. They are also often tied to negative consequences. For example, there is the concept of the "evil eye," a mere look that in cultures across the world is believed to have the power to cause harm or ill fortune. It has a name in many languages—from Arabic to Greek to Hawaiian—and features in the Old Testament. There is also the frequently reported notion of the "unseen gaze," a sense that something or somebody is watching. We also talk of eyes "burning into the backs of our heads," and related perceptions of surveillance, such as the feeling that "these walls have ears." Such concepts are often manifested in religion as well. On the ceilings and walls of some churches, one can find images of disembodied eyes staring down, ever watchful and haunting.

Even where physical eyes are not present, we still have a tendency to feel that important people may be "watching," despite the fact that they might be miles away. We check ourselves as we do things, thinking: what would my wife, or father, or boss say? Even though they are absent, we imagine them watching us, and it can be powerful enough to alter our behavior. The policing can even work with people who are dead. As we have seen, the social role of ancestors is no small matter in many cultures around the world. The same phenomenon persists even in modern secular settings. Although we may not actually believe that there are ghosts or spirits of our ancestors hanging around, it is nevertheless common to find ourselves thinking about what our grandfather might have said. People often do something, or react to something, *as if* their great aunt Agatha is watching (perhaps in horror as we were about to throw away a plate of perfectly good food, or forgot to stand for the national anthem). Ancestors can still haunt the lives of atheists just as they do for Hawaiians or Burkina Fasoans.

A striking feature of agency detection—and one important to this book—is that it has a *negative* bias. Up to now we have been discussing the detection of agency that might be responsible for either good or bad events, but a series of experiments show that we are significantly *more* likely to attribute agency when events are bad. Psychologist Carey Morewedge found that when playing rigged economic games, people tended to suspect the intervention of an agent when outcomes were unfair or represented direct losses, and were more likely to make such attributions than when outcomes were of equivalent size, but represented fair splits or wins. This strongly reinforces the logic of the supernatural punishment theory, and converges with the negativity bias we explored in Chapter 2. People are generally more attentive to, and more affected by, negative rather than positive information and events, and this extends to the detection of agency as well. Even if positive and negative events are equally commonly *encountered*, therefore, we are more likely to believe that the negative ones were *caused* by a supernatural agent.[42]

BELIEF IN A JUST WORLD

And now for the final key characteristic. We have a deeply ingrained sense that people should get what they deserve. And this expectation is so strong that we have a bias to interpret people's dispositions and behavior so that punishments fit the crime. Half a century of research has strongly corroborated the phenomenon. As ethicists Claire Andre and Manuel Velasquez explain:

> The need to see victims as the recipients of their just deserts can be explained by what psychologists call the *Just World Hypothesis*. According to the hypothesis, people have a strong desire or need to believe that the world is an orderly, predictable, and just place, where people get what they deserve. Such a belief plays an important function in our lives since in order to plan our lives or achieve our goals we need to assume that our actions will have predictable consequences. Moreover, when we encounter evidence suggesting that the world is not just, we quickly act to restore justice by helping the victim or we

persuade ourselves that no injustice has occurred. We either lend assistance or we decide that the rape victim must have asked for it, the homeless person is simply lazy, the fallen star must be an adulterer. These attitudes are continually reinforced in the ubiquitous fairy tales, fables, comic books, cop shows and other morality tales of our culture, in which good is always rewarded and evil punished.[43]

Just World theory emerged in the 1970s from experiments conducted by psychologist Melvin Lerner. He found that if students were told another student had won the lottery, they tended to believe that the lucky winner must have been a harder worker than others. Another experiment in which participants observed people getting electric shocks found that the observers reported lower opinions of those receiving punishment—particularly when the victim appeared to be trapped and had no way to escape it. The injustice could only be reconciled by assuming that the recipients must be bad eggs who had brought it upon themselves. These experiments built on a long history of related work in psychology, stemming from the pioneering work of Swiss developmental psychologist Jean Piaget. He noticed that people tend to anticipate or explain people's fates as recompense for prior deeds, and he called this reasoning "immanent justice" ("immanent" as in inherent or intrinsic, not "imminent" as in looming or impending).[44]

What is interesting about Piaget's ideas and Just World beliefs is that it is particularly *negative* experiences that underlie the theory and emerge in empirical tests—misfortunes tending to be attributed to prior misdeeds, rather than positive events being attributed to good deeds. In the decades of research on the topic, expectations and attributions of punishment have been the dominant theme. This need not always be the case, however, and some recent research shows that it can work in the other direction too. For example, psychologist Mitch Callan and colleagues provided vignettes about people who supposedly experienced a car accident or won a lottery, but who had varying histories prior to the event. They found that experimental subjects "viewed the outcomes as the result of prior behavior most when they fit deservingness expectations (good person won the lottery, bad person injured in automobile accident)." The experiment explicitly linked the concepts of Just World beliefs and immanent justice, and found that it held up for both negative and positive attributions.[45]

Other research has found interesting variations on the theme. Psychologist Joanna Anderson and colleagues note that while previous research focused on the fact that "people's need to perceive the world as fair and just leads them to blame and derogate victims of tragedy," perceived injustice can be smoothed out in other ways as well. Rather than people's misfortune resulting from misdeeds, people who had initially high, or primed, Just World beliefs tended to perceive "more meaning and enjoyment in the life of someone who had experienced a tragedy" compared to someone who had not. Thus the need to see the world as just, while usually leading to blaming the victim, can also lead people to see a silver lining in negative events.[46]

Other studies found that in their search for justice, people may engage in behavior that seeks to help them correct a perceived imbalance in their own good and bad fortunes. Subjects who had experienced higher levels of hardship were more likely to express an interest in gambling, and even to gamble excessively. Games of chance appeared to offer an opportunity to redress their luck. A follow-up study, rather than relying on people's self-reported deprivation, gave them varying amounts of money to start with. Those who were primed to feel relatively more deprived (this time by the experimenters) were again more likely to gamble. These subjects also explicitly showed a greater preoccupation with justice and resentfulness. It appears that if people are unable to obtain benefits they feel are due to them, they are more likely to engage in activities that can right the injustice—even if doing so is risky or unlikely.[47]

One might wonder how Just World beliefs are maintained, because our everyday experience is full of *undeserved* events. Why don't we just drop the fantasy as we become older and wiser? Although many events may indeed be undeserved, people seem to be inclined to the idea that in the end, they will be evened out by other events. An experimental study tested this idea and found evidence for a "compensatory bias," an ultimate balancing out of good and bad in the way people interpret events, the way they remember them, and the way they anticipate them in the future. In short, typical Just World studies show that specific events are explained by a specific prior wrongdoing. This study showed that *multiple* events can be parceled up and balanced against other events in the past, present, or future.[48]

But if all of us, including atheists, are subject to Just World beliefs, do they vary among individuals, and in particular are they stronger or weaker among religious believers? One experimental study looked into this, and found that Just World beliefs and immanent justice beliefs are associated with harsher attitudes toward the poor, suggesting that certain people are unmoved by others' misfortune precisely because they are more likely to believe that people are deserving of their fate. However, whether people were religious or not did not make any difference to the relationship. This suggests that personality or other factors may be more important than religion in explaining variation in Just World beliefs, or simply that the Just World effect is powerful enough to pervade people's worldview irrespective of how they may have organized their beliefs around a given religious creed.[49]

Finally, another study asked whether Just World beliefs changed if the scenario explicitly involved religion—here, converting to or from a religion (an event of great significance for believers and atheists alike). In this experiment people read a scenario about someone who had, amongst other information, decided to convert (or not to convert) to a new religion, and then in some of the scenarios this person was subsequently the victim of a "brutal robbery which permanently disabled him." People's responses to this situation strongly depended on religion. They did not readily blame the tragedy on the victim or their morality in general. However, people showed a strong tendency to blame the tragedy on the decision to convert—and they attributed this blame irrespective of whether the decision had been to convert to a new religion *or* not to convert to a new religion. People appeared to believe that accepting or rejecting a religious faith was instrumental in the victim's vulnerability to misfortune, and they made the decision fit the accident whichever way around the logic had to be made given their own personal perspective.[50]

SUPERNATURAL PUNISHMENT: ORDER AMIDST THE CHAOS

The cognitive science of religion tells us a lot about what people tend to believe. What we don't have is a good account of how all these bits and pieces fit together. Sometimes it can seem like a laundry list of psychological dispositions and quirks that have no rhyme or reason, and that we—as humans and scholars—are all over the place. We also lack any consensus on how religious manifestations of these cognitive mechanisms evolved, or even whether they helped or hurt us in our evolutionary past. As I stress throughout the book, I do not think supernatural punishment is a magic bullet that can explain everything about the diversity and complexity of religion. However, it offers an overlooked and powerful *organizing principle*, around which many other aspects fall into place.

People in general—religious and atheist alike—are susceptible to perceptions of cause and effect, mind-body dualism, agency, and Just World beliefs (Table 5.1). Each forms an important part of the scaffolding for beliefs in supernatural consequences of our actions. Table 5.1 also highlights that in three of the four cases, there is a bias that elevates the salience of supernatural punishment over supernatural rewards. First, cause and effect is negatively biased because the goal is to avoid the disaster of failing to identify important sources of danger, even if this means we overestimate them in general. Second, our agency detection is negatively biased because we are more likely to attribute negative events to agency than positive events. Third, Just World beliefs are negatively biased because although they can be about both positive and negative events, it is typically the assignment of misdeeds to *negative* outcomes that maintains people's belief in a Just World. Together, these four cognitive mechanisms converge to push atheists, as well as religious people, into believing that our actions have supernatural as well as material

Table 5.1 Key cognitive mechanisms underlying beliefs in supernatural consequences of our actions, and whether they have positive or negative biases.

	Positive biases	Negative biases
Cause and effect	No	Yes (avoiding costly mistakes)
Mind-body dualism	No	No
Agency detection	No	Yes (negative agency bias; surveillance; evil eye)
Just World beliefs	No	Yes (tends to be misdeeds tied to negative events)

consequences. Furthermore, these beliefs tend to priviledge negative rather than positive consequences—that is, supernatural punishment.

BORN BELIEVERS?

How do we know these phenomena have anything to do with *evolved* human cognition—a result of natural selection rather than socialization? One of the core strands of research in the cognitive science of religion has emerged from developmental psychology and the study of how children develop and learn religious concepts. This is important because it blows apart the argument that superstitious or religious beliefs and tendencies are just learned from one's social and cultural environment. Instead, it seems we are born with these tendencies— "born believers," as psychologist Justin Barrett puts it. The main finding is that children tend to hold beliefs in supernatural causes of events, the afterlife, supernatural agents, and a Just World, from a very young age. As soon as the relevant cognitive machinery is in place (such as theory of mind), children exhibit beliefs in supernatural concepts (such as that there are supernatural agents who know what they know). There are important changes with age, as the brain develops, but children seem to be automatically endowed with supernatural beliefs. Psychologist Deborah Kelemen dubbed children "intuitive theists," because of their teleological reasoning that things exist "for" something (e.g., clouds are for raining). Bering and Bjorklund's study on children's reasoning about the psychological states of dead agents also revealed a default "afterlife" belief that is only replaced by an explicit scientific understanding of biology and death—knowledge that was of course limited in our prescientific evolutionary past. And these intuitions are not morally neutral. It seems that children also have an innate tendency to believe that supernatural agents are especially concerned with right and wrong (again, something that has been explicitly explored and found in experiments). Hence Joseph Bulbulia's argument that "children are not only intuitive theists, they appear to be intuitive supernatural moralists."[51]

In modern societies, the fact that supernatural beliefs are so natural and common in children but then frequently absent among adults suggests a powerful role for social and cultural factors in fostering *atheism* rather than religion—especially western secularism and science education. Atheism must be enculturated, and this may not be easy. Richard Dawkins argues that religion "parasitizes" our underlying cognitive mechanisms and, as a result, cannot easily be shaken later. Whatever the reason for the stickiness of religious ideas, more (secular) education should lead to higher levels of atheism. The empirical evidence on the relationship between levels of education and levels of religiosity, however, is mixed. The World Values Survey shows a correlation between education and atheism in many but not all countries, and other surveys find that within academia atheism varies widely depending on discipline, with many more atheists among social scientists than physical scientists. Nevertheless, psychologist Benjamin Beit-Hallahmi found that "differences among academic field vanished with growing eminence," as religiosity declined among Nobel laureates and other high-flying intellectuals, but it is clearly not a simple relationship.[52]

Sociologist William Bainbridge suggests that atheism has other important causes, such as being raised by atheists, early traumatic experiences with religion, having "resolutely unmystical personalities," a rebellious adolescence, or socialization to antireligious ideologies in one's profession. Atheists tend to be young, male, liberal, and well-educated, a pattern that Bainbridge uses to argue that atheism is a luxury afforded to those who lack social obligations. Other correlates of atheism have emerged from large-sample surveys and experiments. As summarized by psychologist Catherine Caldwell-Harris, atheists are "slightly less social than religious believers, less conformist, and more individualistic. Atheists in particular are overrepresented among scientists and academics, and their high intellectual achievement may stem in part from their preference for logic and their enjoyment of rational reasoning. Lacking interest in a reality beyond this world, nonbelievers focus their moral concerns on social justice and the here-and-now."[53]

The emergence of atheism leaves many questions to explore, especially from an evolutionary perspective—how might skepticism and non-belief have been helped or hindered by natural selection? But the key message is that we—as human beings—seem to be primed to adopt religious beliefs rather than atheistic ones. Atheism has to be learned, but supernatural beliefs are part of human nature. To echo the title of Robert McCauley's book on the topic, "religion is natural and science is not."[54]

CONCLUSIONS

It is a folk tradition in some parts of the world to hang a horseshoe on the wall for luck, open end up, to keep the good luck in. Nobel laureate and physicist Niels Bohr once hosted an American scientist at his home in Tisvilde in Denmark. The visitor was surprised to see a horseshoe hanging over Bohr's desk: "Surely you don't believe the horseshoe will bring you good luck, do you, Professor Bohr? After all, as a scientist . . . " Bohr laughed and reassured his friend, "I believe no such thing, my good friend. Not at all. I am scarcely likely to believe in such foolish nonsense. However, I am told that a horseshoe will bring you good luck whether you believe it or not." Bohr was no doubt joking (although of course someone had nevertheless gone to the trouble of hanging the thing up there), but while we joke about our superstitions, they are pervasive and powerful. Most of us, if not all, whether we are religious or not, harbor a variety of superstitious thoughts or rituals that we find hard to avoid even if we deny their logic or utility. We may not even be aware of them.[55]

Atheists do not expect to be struck down by a thunderbolt when they do something wrong. Beliefs about supernatural rewards and supernatural punishments are clearly much more subtle and variable than that. But the complexities can make it more pervasive rather than less—an unobtrusive tally that builds up in multiple minor ways, or a mindset that only comes to the fore when there are high stakes, high levels of uncertainty, a lack of control, or stress and anxiety. Atheists and believers alike have implicit but powerful expectations that we, and others, will reap or suffer the consequences of our actions, whether from gods, spirits, and ancestors, or from karma, immanent justice, comeuppance, just desserts, and plain old fate. However it happens, we tend to think that what goes around comes around. Concordantly, people are not surprised when those who have done wrong meet a sticky end, a phenomenon so consistent and common it has a name in Just World theory. It seems to be the natural order of things. Our brains bear the marks of a deeply ingrained Big Brother mentality.

Much as religious believers might not accept official doctrine, identify with a specific sect, or believe in a specific supernatural agent, atheists may not strictly be nonbelievers. Psychologists find that human brains are characterized by potent mechanisms that tend to perceive things in the world around us, whether they are there or not—not least causes and effects, mind-body dualism, supernatural agency, and a Just World. Given that we all have the same kind of brain, all of us are susceptible—at least to some extent and at some times—to superstitious or religious thinking. Indeed, discounting supernatural causes and consequences can take real mental effort.

When we explore the rationale behind people's supernatural beliefs, it is often, oddly, *material* evidence that underlies such immaterial beliefs. In Hood's research he finds that "the number one reason given by people who believe in the supernatural is personal experience." People commonly claim to have seen ghosts, for example, or to have experienced coincidences so bizarre that they could not have happened by chance: "For believers, examples of the supernatural are so plentiful, they are impossible to ignore." Remarkably, therefore, this means that it is not intangible *beliefs* themselves that we need to account for, but rather the perception of tangible events—our brains are "seeing" supernatural activity, recording it, and using it to adjust our prior beliefs. This is precisely why they are so convincing, and why they are so resistant to challenge. Having *perceived* something, we have every reason to believe it. As with claims of alien abduction, to many victims, they are for all intents and purposes real. And given that our perceptions arise partly from subconscious processes, they can be beyond our ability to correct or alter them, or even to recognize that they are biased. "Even if you deny having a supersense," Hood warns, "you may still be susceptible to its influence, because the processes that lead to supernatural thinking are not necessarily under conscious or willful control." One might expect that the experience of brute facts in everyday life would teach us to reject false beliefs and become scientific materialists. But in fact the experience itself can become corroborative evidence *against* any objective reality, precisely because our brains evolved to perceive things in certain, biased ways. This sets us up to detect and interpret new events in such a way that they match our pre-existing beliefs. As anthropologist Phyllis Dolhinow once put it, "I would not have seen it if I hadn't believed it."[56]

What we have learned from the prodigious work on the cognitive science of religion is that the brain has a set of cognitive dispositions toward supernatural beliefs that are common to human beings—whatever our culture or creed. Almost inadvertently, cognitive scientists of religion have taught us as much about atheists as they have about believers. It is not that some people are religious and some are non-religious. Rather, *Homo sapiens* as a species—all of us—are inclined to supernatural thinking. In some people this is manifested as religious belief and in others, including atheists, it lingers on in superstitious beliefs and behaviors. The idea that human beings have religious inclinations should not be surprising. I spent a year on a research project with a group of scientists, philosophers, and theologians in Princeton, and one of the few points of consensus that participants from a diversity of fields and approaches agreed on was that religiosity is part of human nature. Darwin himself had noted this, in line with earlier theologians who had recognized more or less the same thing. Sixteenth-century reformation leader John Calvin, for example, spoke of a "sensus divinitatis," and nineteenth-century theologian Friedrich Schleiermacher believed that humans had "a basic religious impulse."[57]

Science has caught up with intuition. In the centuries since, we have developed a remarkable understanding of human biology, brains, and behavior. As it turns out, Darwin, Calvin, and Schleiermacher were right, although they could not have known why at the cognitive and neuroscientific level. Clearly, human beings have a natural tendency to perceive supernatural agency, of which religion is only one example. Human brains are wired to believe that events happen for a reason, and that our actions have consequences. This feeling is pervasive and powerful, churning away even when we are alone and even among atheists trained in statistics and skeptical of coincidences. All of us—believers, agnostics, and atheists alike—worry about unseen eyes observing and judging our actions, even our motives and thoughts. Humans are guided by an inner sense of duty to some kind of Big Brother. It's not just a religious belief. It's bigger than that. It's human nature.[58]

The cognitive science of religion represents a major leap forward in our understanding of religion. However, it carries with it a big theoretical claim, sometimes implicit and sometimes explicit, about the origins of religion: that it is an accident. Since religious beliefs stem so naturally from basic cognitive mechanisms that evolved to serve *other* adaptive functions, religion itself may be a mere byproduct of human evolution. For some researchers, the consistency and strength of these developmental and cognitive propensities is the be-all and end-all of the discussion—human cognition "explains" religion. This is typical of many people working on the topic from a cognitive science perspective. For example, Pascal Boyer's book on the social-cognitive foundations of religion was called "Religion Explained." Many psychologists are serious about this, and researchers like Paul Bloom and Lee Kirkpatrick explicitly argue that God is an accident of humans' big brains. Richard Dawkins makes similar arguments. Religious belief is seen as a spillover of childish thinking that can and perhaps should be eradicated and replaced by science.[59]

But there is an alternative. Just because religious beliefs are a natural *outcome* of underlying human cognitive mechanisms does not mean that they have no influence on our success, survival, and reproduction, or that natural selection will ignore their effects. Given the prominence, diversity, and demands of religious beliefs and behavior, it seems unlikely they would *not* somehow, or at some times, affect peoples' lives and livelihood—especially among the small-scale societies of our past, where supernatural beliefs were universal and powerful. Indeed, supernatural beliefs might have become a highly adaptive trait in the evolution of our species. But how?

CHAPTER 6

GUARDIAN ANGELS

He that does good to another does good also to himself.

—*Seneca*[1]

Spare a thought for the Shaker religion. At the time of writing there are only four Shakers left in the world, all living in the Sabbathday Lake Shaker Village in Maine. Shakers, or more fully, the United Society of Believers in Christ's Second Appearing, originated in 1747 in Manchester, England. Following other persecuted religious movements, they migrated to America, where there has been a community in the hilltop village at Sabbathday Lake since 1783. In a 2006 *Boston Globe* article entitled "Last Ones Standing," reporter Stacey Chase drew attention to the plight of the Shakers, describing the setting around the lake as a "breathtaking pastoral haven of forests, farmland, and apple orchard." Today, "the four faithful live a life of ascetic simplicity and abide by the three C's: celibacy, confession of sin, and communalism." They follow a strict daily routine of meals, prayer, and work, with a day of rest on Sunday. Unlike the Amish in Pennsylvania and Ohio, the Shakers have modern conveniences—phones, the Internet, cars. However, "most activities, from borrowing one of the vehicles to buying a new pair of pants, must be sanctioned by the group, which calls itself a family." This egalitarianism pervades life, and "even in death, the Shakers repudiate individualism. All grave markers were removed from the tidy Sabbathday Lake cemetery in the late 1800s and replaced by a single slab of granite in the center, bearing one word: SHAKERS."[2]

The celibacy clause above may have caught your eye. Shakers aim to imitate the life of Christ, and as Christ was celibate, so are they. Given this constraint, the Shakers' survival relies on converts. In the past, there had been a steady stream of newcomers. At the time Chase visited their community, they were continuing to receive around seventy inquiries a year from people attracted to the Shaker way of life. But after sending out the standard mailing of literature, most of them were never heard from again. Despite the idyllic setting and strong collectivist spirit, the Shakers are slowly, and steadily, vanishing.

The four living Shakers are not unaware of their apparently impending doom. Although "they pray for more converts" the precedent is against them. Other Shaker villages in New England have already fizzled out and "were long ago subdivided into housing lots or turned into prisons." The Sabbathday Lake community is not only threatened by a lack of converts, but also by local development and considerable operating costs. They generate income from their land by leasing cottages, the forests, farmland, and a gravel pit. Plus, it's a tourist destination. Sabbathday Lake draws around 10,000 visitors a year. The Shakers have also been planning to sell rights to Maine Preservation and the New England Forestry Foundation, in an agreement that would preserve the land forever. The village may therefore persist—at least as a museum—even if the villagers are gone.

But if they are to survive, who might still join them? And how? Becoming a Shaker is no easy business. Entrance requirements are daunting. Converts cannot be married, cannot have children, and cannot have any debts. Only after a full year of laboring in the community does the Shaker group decide whether the newcomer is up to scratch. And not for another *five* years would they acquire the status of a full community member, which demands, on top of everything else, the "consecration of one's worldly goods to the communal sect." With such barriers to entry, it is perhaps not surprising that the population has dwindled and the Shakers have become an increasing rarity at Sabbathday Lake. One of the Shakers recalls a tour guide saying to a group of visitors: "This is not a whale watch. We neither guarantee nor will you be likely to see a Shaker on the tour." But like many whale species, it seems, the Shakers are in real danger of extinction.

The Shaker story has a curious twist involving the very reporter quoted above, Stacey Chase who wrote the 2006 *Boston Globe* article.

No, she did not become a Shaker; it's even more bizarre than that. The youngest of the Shakers, and therefore the likely candidate to have become the last one on Earth, Brother Wayne Smith, renounced the religion and his twenty-six-year vow of celibacy, and left the community to marry Chase! He had been, she noted, "one of the most unattainable men on the planet." She recognized the irony, and later reported matter-of-factly that, "his sudden departure not only shocked the three remaining Shakers, but it was as if it validated the protracted, public deathwatch of the Shaker faith."[3]

HOW COULD COSTLY RELIGIOUS BELIEFS EVOLVE?

Many people see religion as a puzzle: Religious beliefs and behaviors demand a significant investment of effort, time, and resources that could be spent on other things, and the benefits are not always obvious. The economics of religion, from a purely material perspective, doesn't seem to add up. An evolutionary perspective makes the puzzle even more perplexing, because natural selection tends to stamp out costly traits. Individuals who invest in costly activities that do not advance their own survival and reproduction generally disappear. The idea that an organism might voluntarily adopt an explicit strategy of celibacy, like the Shakers, takes the puzzle to even greater depths. How can we explain the origins and persistence of such religious beliefs and practices?

First, we should recognize that not everything humans do is necessarily functional. This has always been the case, but is especially true today. There are numerous contemporary cultural phenomena that have little or no systematic impact on Darwinian fitness (survival and reproductive success), from art and music to fashion and cricket. All of these may have *roots* in adaptive aspects human social behavior, but are not necessarily adaptive in themselves today. Natural selection does not wipe out people who like tacky art or wear badly cut jeans, much as fashion fascists might like. And so it may be for religions.

Even if religious beliefs and behaviors have adaptive *origins*, we need not expect every aspect of every religion to be adaptive *today*. Some cultural traits are neutral, neither helping nor hurting the individuals or societies that bear them, just as biological traits that have little bearing on survival and reproduction can crop up or stick around—such as our wisdom teeth or appendix. But other cultural traits can be systematically *detrimental* to fitness—maladaptive—and sometimes so much so that they drive their bearers to extinction. Jared Diamond's book *Collapse* describes a number of societies that died out because of the persistence of harmful beliefs or practices (which may have been adaptive at other times or places), from the Easter Islanders who cut down every tree, to the Greenland Norse who didn't like to eat fish. Perhaps certain Shaker proscriptions are just maladaptive manifestations of religious beliefs and taboos—which in general are highly adaptive. Most religions do not demand celibacy. It is a problem for the Shakers but not for religion in general.[4]

Still, ruling out reproduction is rather extreme. It is common for particular individuals *within* a religion to be celibate, such as priests, monks, or ascetics, and these individuals (and their behavior) may in fact serve some greater function in the society as a whole. But a celibacy constraint for an entire group seems to sound an evolutionary death knell. It offers a vivid example of how a religious belief or behavior can damage Darwinian fitness. Many other, less drastic beliefs and practices can *act against* Darwinian fitness, if not quite so dramatically. The effort, time, resources, and missed opportunities involved in following religious commitments, performing religious rituals, sacrificing precious food, or abstaining from profitable activities can all harm fitness. Celibacy may be rare, but these other practices are quite common. Thus even if we set aside cases of the denial of reproduction, to understand religion in general we still face a significant paradox: How could costly beliefs evolve?

A belief that we are watched and punished by gods or spirits is a case in point, given that these beliefs seem to directly curtail the self-interest of the individual. Why believe in something—especially something that stems from invisible agents that no one can see—if it chips away at our own chances of survival and reproduction? How could such a belief ever get through the mill of natural selection?

SINNERS AND WINNERS: IS IT INDIVIDUALS OR GROUPS THAT BENEFIT FROM BELIEF?

The most obvious solution to the paradox is that, while a belief in supernatural punishment may reduce individual self-interest, it increases cooperation in the group or population as a whole, which would seem to be good for everyone. However, the argument that a belief in supernatural punishment (or indeed anything else) evolved *because it is good for the group* doesn't cut it with conventional evolutionary theory. It invokes the problematic notion of group selection that we encountered as part of Darwin's puzzle in Chapter 2. If I make sacrifices for the good of the group, then other individuals who do *not* make such sacrifices will do better than me and spread at my expense. Altruists will die out and selfish individuals will prosper. That's the logic of natural selection, and it can be found throughout nature. Trees would be better off if they grew slightly shorter (saving resources). Baboons would be better off if they did not use their dangerous canines to fight (reducing injuries). And humans societies would be better off if the rich gave their money to the poor (spreading the wealth). All such idyllic scenarios would be good for the group, and fruitful for the broader population or species, but they all suffer from the same problem. They would be wide open to exploitation by selfish individuals—trees that grew a bit taller than the others, baboons who could win fights with a better weapon, and humans who exploited others' generosity. These more selfish individuals would do better than everyone else, have more offspring, and their exploitative traits would spread. Selection favors relatively successful individuals over relatively unsuccessful individuals—regardless of what is best for the group.

Group selection has thus traditionally been seen as a flawed idea, because selection is thought to act primarily at the level of individuals, not groups—after all, it is individuals who succeed or die (or, strictly speaking, genes), and are thus selected or not selected by natural selection. In recent years some evolutionary biologists, following the lead of John Maynard Smith and David Sloan Wilson, have argued that group selection may in fact operate in certain circumstances. Although individuals who sacrifice for the good of the group may suffer reduced reproductive success themselves, if the group as a whole does better *than other groups* as a result, then the more altruistic group—and the sacrificial genes within it—could spread at the expense of groups composed of more selfish individuals (perhaps because, for example, altruistic groups make more effective warriors or competitors than selfish groups). It remains a theoretical possibility, and one that if true, would make the supernatural punishment theory especially simple and powerful (belief in supernatural punishment suppresses self-interest and makes for more cooperative groups, which are thereby better able to outcompete groups composed of more selfish skeptics; belief in supernatural punishment therefore spreads and strengthens). That would be pretty straightforward, and we could end the book here.[5]

However, group selection remains controversial, and empirical examples in humans and animals are few. Biologists therefore question whether the special conditions that would be necessary for selection to operate among groups—even if possible in theory—are ever present in nature, or ever strong enough to work. And they would have to be strong, because individual selection will be churning away as normal and tends to work in the opposite direction. What matters is the strength of selection operating *at the level of groups* compared to the strength of selection operating *at the level of individuals*. To win the tug of war, any selection acting at the group level would need to be stronger than the typically fierce selection acting on individuals.[6]

The debate rages on, but the argument of this book glides over it, for two reasons. First, because you do not need group selection for beliefs in supernatural punishment to evolve—I argue in this chapter that they arise instead out of benefits to the *individual*. Second, because in this case individual and group selection would work in the *same* direction—beliefs in supernatural punishment would help individuals and groups alike. If there is no group selection, beliefs in supernatural punishment will be selected for among individuals. If there *is* group selection (whether operating on genetic or cultural traits)—beliefs in supernatural punishment would merely be selected for even more strongly and rapidly.

So what is it that makes belief in supernatural punishment an advantage to us as individuals? This is where it gets interesting. Believing you are being watched and potentially punished by supernatural agents would seem at first glance to work against us. But it can, in fact, *promote our own self-interest.* The critical evolutionary question is what's in it for the individual—why would the genes of individuals who curb their activities because of invisible supernatural agents survive and spread better than the genes of individuals who reject or ignore such agents and do what they want? The answer is that selfish behavior, in our unique surroundings of cognitively sophisticated and thus socially transparent groups, carries its own costs that we may do well to avoid. The gains from selfish behavior, in other words, can be trumped by the costs of reputational damage and retaliation. Our self-interest may be better served by following social rules and acting well in the eyes of others than by doggedly pursuing our own material gains. But dispensing with our deeply wired self-interest is easier said than done. The prospect of our actions being constantly observed and potentially punished by ever-present gods helps to avoid being—and avoid appearing to others as—overly selfish, egoistic, greedy, lazy, offensive, domineering, or otherwise in violation of social norms and taboos. This chapter therefore explains a crucial argument at the heart of the book—how believing in supernatural punishment can increase Darwinian fitness for the individual believer. If so, beliefs in supernatural punishment would have been favored by natural selection over the course of human evolutionary history.

SELFISHNESS BECOMES MALADAPTIVE

From the beginnings of life on Earth, genes that are better at replicating copies of themselves are more likely to persist and spread, and they thus tend to give rise to traits that are egoistic. Of course, there are complications and caveats—Darwin himself noted widespread instances of self-sacrifice in nature—but a century of research has shown that such apparent self-sacrifice is only selfish genes' way of propagating themselves more effectively by helping kin, reciprocating favors, or signaling generosity to attract mates or allies. After 3.5 billion years of evolutionary history, however, the human lineage ran into a new and significant danger created by our own selfish behavior. Life changed for us the moment that we evolved two unique cognitive abilities: (1) a sophisticated "theory of mind" (our ability to imagine what other people are thinking, to know that they know things, and to recognize their intentions); and (2) complex language (our ability to transmit and discuss abstract information with others). No other species on the planet have these cognitive abilities. Chimpanzees and some other high-intelligence mammals such as dolphins may have a rudimentary theory of mind, but their level of sophistication is very basic compared to our own. And, although many animals have some form of language, broadly defined, they have nothing like our *complex* language that is able to deal with abstract concepts and ideas.[7]

The reason theory of mind and language are so important is because of their implications for *selfish* behavior. Suppose Tom has sex with Harry's wife Sue while Harry is away on a hunting expedition. If they were chimpanzees, Harry would never know—Tom's self-interested behavior can pay rich evolutionary dividends. By contrast, if they are both humans, Harry is inordinately more likely to find out, one way or another. Harry, or someone else, might well figure out or learn what happened (Tom or Sue might even let it slip themselves). Harry can also exploit language to gain information, interrogate the perpetrators and witnesses, trick people into confession, spread rumors, recruit allies, and seek revenge. Adultery is far more risky for modern humans than it was for our ancestors. The same logic applies to numerous other self-interested behaviors. Today, of course, clandestine behavior tends to go on behind closed doors, which may help to conceal it. But the period of interest for our developing evolutionary theory is the time before doors (and walls) were invented. In small-scale hunter-gatherer societies privacy was not only difficult to attain, but was rarely sought, and was frowned upon when it was. As evolutionary economist Anna Dreber and her colleagues put it, because we have evolved over millions of years in small groups, "our strategic instincts have been evolving in situations where it is likely that others either directly observe my actions or eventually find out about them." The means of tracking others' endeavors can be surprisingly effective, even if indirect and perhaps unfamiliar to us in modern, western society. For example, one study of the Mehinacu Indians in central Brazil found that they could draw accurate footprints of specific individuals in the village. People left a map of their activities wherever they went. In a variety of ways, actions do not have to be directly observed to be suspected or discovered—but it does require brain power.[8]

The implications of this are vast. In our socially transparent society, a society now brimming with language and theory of mind, the *same* selfish desires that had always been essential to genetic survival could lead us into enormous trouble. And they continue to do so. Pick up any newspaper and you will find plentiful examples of scandal and greed, and if it's in the news then those involved are usually paying the costs as well as reaping any profits. We also find evidence in our universal and incessant efforts to counter the problem. As psychologist David Barash has written, "people are widely urged to be kind, moral, altruistic, and so forth, which suggests that they are basically less kind, moral, altruistic, etc., than is desired." There was, therefore, a defining moment at some point in our evolutionary history where the entire cost-benefit calculus of selfishness was more or less thrown out the window. Selection no longer operated on us in quite the same way it did on other animals. In a clever and gossiping species, knowledge of selfish actions could spread and come back to haunt us with a vengeance. Selfishness used to help, but now it could hurt.[9]

The problem can be summarized with a snapshot comparison of life before and after the advent of our cognitively sophisticated big brains. *Before* the evolution of theory of mind and complex language, selfish behavior would be consistently favored by natural selection as long as its benefits for fitness outweighed the costs (even if these behaviors occurred in full view of other individuals, since they could not tell anyone else). For example, chimpanzees can rape and steal in front of other chimpanzees without their behavior being discovered by or reported to others who might do something about it but are absent—namely, the alpha male of the group, or the allies or relatives of the victim. As evolutionary anthropologist Robin Dunbar wrote, "for monkeys and apes, all this has to be done by direct observation." There can therefore be *no negative repercussions from absent third parties*. Absent individuals could not entertain others' knowledge states, nor could they learn such complex information by communication. Sherlock Holmes would have little to do in chimp society: He would have no interviewees, and he would have no concept of witnesses or motive.[10]

After the evolution of theory of mind and complex language, by contrast, it was now in our genes' interests to *avoid* selfish behavior in many contexts where there could be negative repercussions for fitness. Now—for the first time in evolutionary history—one had to worry about reprisals from other actors, wholly removed from the scene of the crime. People could hear, discover, infer, remember, report, hypothesize, plan, and act on others' behavior—even long after the event. Negative consequences for selfish behavior therefore became both *more likely* (because selfish acts were more likely to be discovered), and *more severe* (because selfish acts could be penalized by groups of people rather than just individuals). But what exactly were these negative consequences?

RETALIATION AND REPUTATION: THE HUMAN COSTS OF SELFISH GENES

Selfish behavior could be costly for two different but related reasons: the danger of *retaliation*, and the damage to one's *reputation*. These prospects are similar in that they result from the ire of fellow group members, decrease prospects for future cooperation, and decrease Darwinian fitness. But the way they incur these costs is different.

RETALIATION: PUNISHMENT, SANCTIONS, AND DENIAL

Selfish behavior can incur direct costs through retaliation. This can have at least three different forms: (1) physical *punishment* (such as pain, injury, or death); (2) material *sanctions* (such as fines, confiscation of property, or ostracization); and (3) the *denial* of benefits (such as shunning, exclusion, or detention). Although the first may seem the most significant, in the small-scale societies of our evolutionary past, all three forms of retaliation could significantly damage one's Darwinian fitness and the number, health, or prospects of one's offspring.

Interestingly, theory of mind and language raised the *severity* of retaliation as well as its likelihood, because these same cognitive capacities increased the number of people who could plan and participate in retaliation. Ten people learning of a misdemeanor and attacking, sanctioning, or refusing to help the perpetrator is much worse than a single individual doing so. Retributive gangs also enjoyed safety in numbers. Group punishment thus became especially costly for the victim at the same time as becoming especially cheap for the punishers, partially solving the second-order free-rider problem we met in Chapter 2 (which was, basically, who's going to step up to carry out punishment given its costs?). People will be much more willing to punish if doing so is cheap yet powerful, and it will be much more effective as well.[11]

Chris Boehm argues in his book *Moral Origins* that group punishment was an important means of deterring domineering individuals as well as cheats, who were likely to be kicked out of the tribe or even killed if they threatened the welfare of others. Even weak individuals could group together to overpower or overrule strong individuals. But such punitive action may not always have been possible or even desirable. For one thing, even if carried out in groups, punishment still carries some cost in terms of time, energy, resources, missed opportunities, and the danger of counter-retaliation (the victim might assemble a gang of his own). And perhaps worse, if group members themselves are the arbiters of punishment, what may seem to be a restoration of justice for one side can simply alienate or lose a potentially valuable member of the (usually small) group, or descend into cycles of violence and blood feuds, all of which may reduce Darwinian fitness for one or both sides.[12]

Direct physical punishment among group members for infringements may not, therefore, have been the most important cost of selfishness. Ethnographic studies suggest that among small-scale hunter-gatherer societies, there is a relative *absence* of direct punishment. Punishment does occur, and it can sometimes be severe, but in general it is rather tempered compared to the widespread punishment we see in modern, historical, and ancient civilizations.[13]

But while direct physical *aggression* may be rare, sanctions and denial are common. What matters for the supernatural punishment theory is whether there is a significant cost imposed on transgressors, however it is imposed. In ancient Hawaii, for example, sanctions and denial could be just as significant as direct punishment: "There were traditional Polynesian ways of handling behavior problems and family disputes. A youngster who misbehaved and refused to conform would receive no further attention, but would be completely ignored, as if he or she did not exist. No word of response, no food or place to sleep could be offered to someone who did not exist. Such tentative banishment usually forced a quick return to acceptable behavior, for the loss of one's place within the clan meant loss of all benefits and protection. In Ancient Hawai'i, ostracism could be a death sentence."[14]

When retaliation for transgressions does occur in small-scale societies, it is often carefully discussed and orchestrated by the community so as to avoid escalation. In many cases, transgressions are punished by the payment of resources, such as animals, to the offended party. In such societies this is no small matter, because animals are not just valuable sources of food. They represent status, bargaining power, and opportunities for attracting husbands or wives into the family. The fact that punishments can involve lengthy negotiations and mediation by important members of the community is also a sign of its significance.

Social mechanisms for resolving conflict are sought precisely because without them victims or vigilantes may take punishment into their own hands, with often detrimental effects for the group. Victims or relatives of those wronged, who are most likely to seek punishment, often have to be restrained or artfully persuaded not to act, and sometimes this fails. Aggrieved victims are not easy to calm and perpetrators are not easily reprimanded. The ramifications of a single conflict can drag in a large network of families, allies, and enemies, which may split the group apart and last for generations.[15]

Across the world's indigenous societies, when infractions are grave enough, transgressors may find themselves at risk of exclusion, banishment, or harm. Selfishness may still pay dividends some of the time, or even most of the time, but in our socially transparent society it can carry significant costs and occasionally even the ultimate price. The danger of pursuing self-interested behaviors that are likely to offend others—for whatever reason—may not, therefore, be a risk worth taking. Even if the group does not collectively or consistently mete out punishment, there is always the possibility that *someone* will be motivated to retaliate. As Ice Cube captured the logic in a 1993 rap song, "check yo self before you wreck yo self / cause shotgun bullets are bad for your health."

REPUTATION: LOSING OPPORTUNITIES

Even if none of the three forms of retaliation—physical punishment, sanctions, or denial—are carried out, selfish behavior can still incur costs. As Karl Sigmund wrote in a review highlighting the important role of punishment in human cooperation, "in small-scale societies, or village life, reputation might have a more pervasive role. It is easier to gossip behind the back of a bully than to confront him." This may sound like a rather weak deterrent. In fact, however, and especially in small-scale societies, reputation is extremely important. In hunter-gatherer groups similar to those in which we evolved, there is no formal authority and few means to dominate or coerce other members of the band. Instead, one has to nurture a good reputation to attract critical resources such as cooperation, help, mates, friends, and allies. Success cannot be won by brute force, and people can simply avoid free-riders and cheats and cooperate with someone else instead. Maintaining one's reputation, therefore, is crucial to Darwinian fitness. Moreover, reputation can take years to build but be lost in an instant, and it can be hard to repair as well. Now that you know Jim broke your trust once, how do you know he won't do it again? The negativity bias of Chapter 2 rears its head again here, making reputational *damage* a particular problem. When experimental subjects were asked how many lives a murderer would need to save—on separate occasions, always putting his own life at risk—to be forgiven for his crime, the median answer was twenty-five. As the authors concluded, "two rights don't make up for a wrong."[16]

One might think that mere gossip—second-hand information—is not likely to be a strong influence on people's decisions about with whom and how much to cooperate. We can make our own mind up about who we interact with. But in fact it may be rather powerful. Dunbar goes so far as to suggest that social information was of such importance that language itself evolved "to allow us to gossip." Gossip may even serve as a *more* dangerous weapon than first-hand information (from both the victim's and the perpetrator's perspective). First-hand experience and direct interactions with people can be valuable—getting our measure of their loyalty and trustworthiness from the horse's mouth. However, second-hand gossip can reveal things that would otherwise be impossible to discover from normal interactions. For example, person A might be able to tell you how person B spoke or behaved in a rare event (such as during a crisis or an opportunity for exploitation), in a certain kind of social context (such as with an important ally or enemy), or simply when you were not there—after all it is precisely how Person B thinks and acts *in your absence* that they might want to hide from you. In addition, there is simply more data from multiple other people's testimony than the few interactions you may have experienced with a given person yourself.[17]

Often the most damning information comes from intercepted gossip—eavesdropping. Person A may not *tell* you everything person B said or did, but you may overhear it. Linguist John Locke argues that eavesdropping itself is a universal human behavior with deep evolutionary roots. Our senses are always alert to information, whether it was intended for us or not, and this offers a method of obtaining strategic information that may have significant consequences for survival and reproduction. Observation of and listening to and about others is of profound importance. And interestingly we often can't wait to pass on any tidbits we may have gleaned. The more juicy the gossip, the more we are compelled to tell others, and the more attention we give it—as we often find, "our ears are burning."[18]

For all these reasons, gossip makes reputations particularly vulnerable, and also therefore particularly *valuable* for deciding how to allocate time and resources among potential interaction partners. A series of theoretical models, laboratory experiments, and fieldwork attest to a vital role of maintaining good standing and prestige in the evolution of human cooperation. Given the importance of reputation, selection pressures acting on mechanisms to preserve it—not least a fear of supernatural observation and punishment—are likely to be strong.[19]

Laboratory experiments have aimed to test some of these ideas. For example, psychologists Jared Piazza and Jesse Bering had subjects play the dictator game that we learned about in Chapter 3, where people decide whether or not to give money away to a stranger. Half of the subjects were told that the recipient would discuss their decision with someone else—this was the "threat of gossip condition"—and for some of this half, that third party was someone to whom they had disclosed personally identifying information. What they found was that prosocial behavior (giving generously) was indeed higher when there was the possibility of gossip, but only if the third party knew their identity. When people could remain anonymous, they were much more self-interested. The experiment therefore lent support to the idea that people's generosity is oiled by threats to their reputation. It had already been demonstrated that people behave more cooperatively in the presence of observation, face-to-face contact, or with an audience. What Piazza and Bering showed was that this effect could be traced to the danger of reputational damage.[20]

Clearly, we do not need to rely on retaliation alone for Darwinian fitness costs to be imposed on cheats and free-riders. Damage to their reputation, and the consequent reduction in the number and quality of willing exchange partners, allies, helpers, and mating partners, could be just as injurious to Darwinian fitness as retaliation. Simple gossip was a significant threat. And it is not only deeds that could spread gossip and endanger reputation, but merely spoken opinions, intentions, and slander as well. People had to be careful about what they said as well as about what they did. This danger is widely recognized in indigenous societies. It was said in ancient Hawai'i, for example, that "a word thrown as a spear may fly back and slay the speaker."[21]

There is much we do not know about our distant evolutionary past, which often leads to speculation. But there is something that we do know happened at some point or other for sure: As soon as we evolved theory of mind and complex language, *selfishness became significantly more costly than at any previous time in our evolutionary history.* And these costs could arise from both retaliation and reputational damage. In the currency of Darwinian fitness, both are bad for business. In our newly transparent society of prying eyes, ears, and minds, egoistic behavior had become a liability. While selfish behaviors might have paid off in the simpler social life of our evolutionary forebears, many of them (or too many of them) would bring a net fitness cost in a cognitively sophisticated, gossiping society. How would we adapt to survive and thrive in such a world?

EVOLUTION STRIKES BACK: ADAPTATIONS TO THE NEW COSTS OF SELFISHNESS

It is commonly thought, as Michael Ghiselin once put it, that "if natural selection is both sufficient and true, it is impossible for a genuinely disinterested or altruistic behavior pattern to evolve." In fact, this is not quite as straightforward as it seems. If selfishness *itself* becomes costly, then it will act against Darwinian fitness. Individuals may, therefore, better serve their self-interest by being less selfish!

Evolution responds rapidly to eliminate costly behavior, so we can expect that if selfishness began to damage reproductive success then natural selection would have favored corrective mechanisms that guarded against the committing, discovery, or punishment of selfish behavior. As soon as we evolved theory of mind and complex language, this became an important problem to solve. If I continued behaving like a chimpanzee in this new social world, while you modified your behavior to avoid the wrath of absent others, people like you would have left more descendants than people like me. This is a remarkable twist in human evolution, but does not conflict with traditional views of natural selection as long as we are careful to distinguish the self-interest of *individuals* from the self-interest of *genes*. As Richard Dawkins explains, "the position I have always adopted is that much of animal nature is indeed altruistic, cooperative, and even attended by benevolent subjective emotions, but that this follows from, rather than contradicts, selfishness at the genetic level. Animals are sometimes nice and sometimes nasty, since either can suit the self-interest of genes at different times. That is precisely the reason for speaking of 'the selfish gene' rather than, say, 'the selfish chimpanzee.'" So what we are concerned with here is figuring out how genes may have responded to the fitness damage incurred by their blundering human vehicles, as they trundled out into a dangerous new social landscape.[22]

However, it is tricky to figure out what that corrective mechanism was. Simply reducing our overall level of selfishness may not have been effective or even possible, because the evolutionarily ancient mechanisms giving rise to selfish behavior cannot be easily thwarted. Basic desires such as sex, hunger, and dominance have been wired into our neural and hormonal system over many millions of years, long before we developed into *Homo sapiens*. Indeed, physiological causes of these desires stem from the limbic system, the ancient part of the brain that we share with pigs, rats, and lizards. The cool-headed rational calculations that can restrain selfish desires, by contrast, come from the neocortex—the modern part of the brain that has developed much more recently. Although each of these brain areas are complex and interconnected they tend to offer broadly conflicting recommendations for action, and it is hard to know which one will prevail, especially in heated or emotional situations. Often, our emotions get the better of us. What is more, the limbic responses that drive our brain chemistry, hormones, and behavior remain powerful and fitness-enhancing (we never need reminding, for example, to eat, have sex, or defend ourselves). They are not only physiologically entrenched but are highly adaptive and sometimes life-saving (or offspring promoting). We cannot simply shut these proximate mechanisms down when it is convenient. It is hard to do, and potentially counterproductive. So what could we do instead?[23]

Evolutionary psychologists Jesse Bering and Todd Shackelford suggested that theory of mind opened the way for the natural selection of new traits that militated against the public exposure of transgressions (rather than against carrying out the offending behavior itself). They focused on traits that could "rescue" inclusive fitness *after* the individual committed some selfish act (e.g., cognitive processes underlying confession, deception, manipulation, blackmail, killing witnesses, suicide, and so forth). The idea that natural selection could operate on such complex premeditated behaviors was a major revelation. However, the particular traits that they examined are posthoc damage-limitation exercises, which are themselves potentially costly, may not work, and might even get us into deeper trouble than we already were. In Dostoyevsky's *Crime and Punishment*, when destitute student Rodion Raskolnikov is surprised by the sister of his victim at the scene of the crime, he kills her too. He eliminates a witness to his crime, but now he could be punished for two murders rather than one. Natural selection might, therefore, favor similarly complex traits, but ones that *constrain* selfishness in the first place. There is, in other words, a more positive side to Bering and Shakelford's story. And there is some evidence too. We see such traits in human interactions every day—restraint, self-control, sacrifice, sharing, patience, and so on. People who carried on being indiscriminately selfish could be outcompeted by prudent others who were able to inhibit their more ancient selfish motives and refrain from breaching social rules to begin with. That's what most people, most of the time, seem to be able to do anyway.[24]

If violations are committed, relying on covering them up, and keeping them covered up, has many traps. There is an arms race of deception and detection, and one hears examples all the time about one winning out over the other. But it seems that our acting may be generally less convincing than our skills at detection. Cheaters are losing the arms race. Psychologist Erving Goffman, in his remarkably penetrating analysis *The Presentation of Self in Everyday Life*, suggests that "the arts of piercing an individual's efforts at calculated unintentionality seem better developed than our capacity to manipulate our own behavior, so that regardless of how many steps have occurred in the information game, the witness is likely to have the advantage over the actor."[25]

Natural selection does at least seem to have worked to make us aware of the impact of our selfish behavior on others. For example, we have developed adaptations to allow us to empathize with and, to some extent, even *feel* the experiences of other people. In recent years neuroscientists such as Marco Iacoboni at the University of California, Los Angeles, have discovered so-called mirror neurons. These remarkable brain cells are activated when we experience a particular emotion, *but are also activated merely when we observe others experiencing the same emotion*. Moreover, mirror neurons are more advanced in humans than their counterparts in animals, suggesting that they have been under further selection pressure from the specific demands of our own social environment, and have become especially adept at taking advantage of theory of mind and abstract thinking. As Iacoboni put it, "neural mirroring solves the 'problem of other minds' (how we can access and understand the minds of others) and makes intersubjectivity possible, thus facilitating social behavior." The biology is remarkable, but once we *have* this heightened awareness of others' mental states and the likely social consequences of our actions, what do we do about it?[26]

What we need is not something to reduce our (essential and adaptive) selfish desires. That would be like putting a limiter on a racing car purely to avoid occasional crashes on corners—it may avoid the problem, but would never be a winning strategy. Instead, we need a mechanism that makes us more prudent about *when to follow* our selfish desires, and *when to suppress them*. We need better brakes, not a slower engine. It needs to be strong enough to counter passions and urges, and it needs to be sensitive to context. Whatever that mechanism is, it has obviously been partially successful—for example, we do not have sex in public places, gorge on food during job interviews, or attack our bosses when they insult us. At least not most of the time. The important point, however, is that there are often failures as well. There are many people languishing in prison for crimes of passion, for example, and most of us can recall social gaffes where our limbic system rode roughshod over someone else's neocortical sensibilities. When humans acquired theory of mind and language, they were thrust into a social minefield. As carefully as we may believe ourselves to tread, there is always the possibility that we will take too great a risk and make a mistake. Sometimes, these are minor missteps that are of little consequence for others or ourselves—we can retreat from danger or defuse a potential problem. But at other times they can be major blunders that bring threats to life and limb. It seems that we would benefit from as much help as we can get in navigating this dangerous environment.[27]

David Barash suggests that "the real test of our humanity might be whether we are willing, at least on occasion, to say no to our 'natural' inclinations, thereby refusing to go along with our selfish genes. To my knowledge, no other animal species is capable of doing that." We can do it. But we may need help, and religion seems to have been one of the helping hands we had along the way. Gods offer a way to reform the selfish ape inside us so that we can avoid the wrath of the hairless and brainy apes around us. People who developed a *fear of supernatural punishment* for social transgressions may have experienced lower real-world fitness costs than more indiscriminately selfish individuals who did not care

Figure 6.1 "The Temptation of St. Anthony," Paul Cezanne (c. 1870). In the course of St. Anthony's wanderings as a hermit in the Libyan desert, he encountered all manner of temptations but was able to refuse them all due to his religious convictions. Courtesy of the Bührle Foundation, Zürich, Switzerland.

what God (or some other supernatural entity) thought of their actions. We may not all have the fortitude of St. Anthony, whose faith enabled him to resist a series of temptations in his pilgrimage in the desert (Figure 6.1). But it does not have to work perfectly, and we do not have to be saints, to benefit from its cautionary effect. Fearing the wrath of God makes us internalize and consider the consequences of our actions especially carefully—*even when we think we are alone.* After all, He is always watching. As Sam Harris has noted, "the very idea of privacy is incompatible with the existence of God." While the average human, therefore, may have an evolutionarily ancient devil on one shoulder prompting egoistic behaviors, we also have an evolutionarily recent guardian angel on the other shoulder: an adaptive inclination to fear

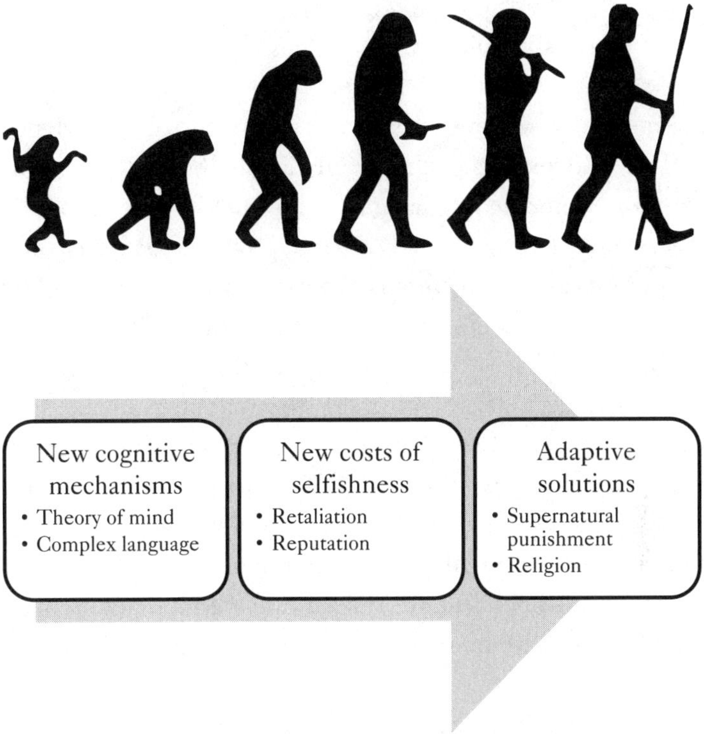

Figure 6.2 As we developed into modern humans, selfish actions became increasingly costly because they were more likely to be discovered (due to the evolution of theory of mind and complex language), and the consequences were more likely to be severe (due to retaliation and reputational damage). A fear of supernatural punishment reduces selfish behavior and its costly consequences, and became an especially powerful deterrent when institutionalized within a shared religious narrative. © imageZebra/ Shutterstock

the consequences of our actions. And fearing *supernatural* consequences can be a more effective deterrent than the weaker and fallible threats of mere mortals. Evolution may thus have favored beliefs in supernatural punishment, because individuals that held them were more successful than peers who did not (Figure 6.2).[28]

VIRTUE AND VICE: WINNING
THE INNER STRUGGLE

Are our instinctive desires really so hard to control that we need the wrath of gods to keep us in check? Obviously there is a lot of individual variation. Some people have a remarkable degree of self-control and others have a remarkable lack of it. Psychologists have come up with ways to measure our impulse control both through experiments and surveys, and they find it varies considerably. Studies have also shown aspects of hard-wiring that affect the ability to control our own behavior, such as neurological differences in people's control of anger, and genetic variation in people's reactions to provocation. It can also reach extreme levels. There are psychiatric conditions, such as Tourette's syndrome, in which the ability to control socially acceptable behavior breaks down completely. But on the whole, most people find that there are indeed many domains and many settings in which yes, we do need help in constraining self-interested or anti-social behavior.[29]

First of all, most of childhood is characterized by a persistent inclination to break social rules. It takes about two decades for people to learn how to control their impulses and desires—with, frankly, mixed results. Second, even among supposedly socialized adults, many of us continue to fail to control our behavior—at least at some times or in some situations. As Bruce Hood observes, "sometimes our capacity to be reasonable is undermined by our gut reactions, which can kick in so fast that it's hard to rein them in with reason." Perhaps the majority of people are able to take this in stride and fight the internal battles between the opposing incentives surrounding health, food, drink, exercise, work, play, spending, saving, and so on. But nearly everyone has experienced it. Some have to seek help. For example, people sign up for dieting classes, alcoholics anonymous groups, marriage counseling, and anger management. They go on training courses to improve their skills in dealing with family, friends, clients, bosses, and workers. Others join talk-it-out classes for all manner of social challenges. It is said that people often join

AA groups precisely when they give up believing that *they themselves* can control their habits. Personal battles are less obvious than these organized interventions, but they are no less pervasive. The struggle for self-control appears to be a universal feature of human nature, and one rooted in our psychology—indeed, in the way the human brain works.[30]

OF TWO MINDS: THE NEUROSCIENCE OF CONTROL

In a famous article that has come to underlie a lot of modern thinking in evolutionary biology, Nobel Laureate Niko Tinbergen warned that "the human brain, the finest life-preserving device created by evolution, has made our species so successful in mastering the outside world that it suddenly finds itself taken off guard. One could say that our cortex and our brainstem (our 'reason' and our 'instincts') are at loggerheads. Together they have created a new social environment in which, rather than ensuring our survival, they are about to do the opposite. The brain finds itself seriously threatened by an enemy of its own making. It is its own enemy." His point was about the danger of human instincts going awry amidst the technology of modern war. But the underlying observation that our brain can act as its own worst enemy seems to have been a problem for a long time.[31]

Scholars over the centuries have intuitively noted that the brain can give rise to conflicting thoughts, and in recent years psychologists and neuroscientists have shed light on why that might be literally true. The root of the problem seems to be evolutionary history. When evolution makes things, it must build them up in blocks. Once a block is in place, other blocks must be laid on top of what is there already. Facing a new challenge, evolution cannot knock down its Lego tower and come up with an entirely new structure. It has to work with what it has. And so it was with the brain. We share much of our brain structure with other animals—for example the limbic system, which includes the hippocampus and amygdala (responsible for encoding memories and processing reward and fear responses). What distinguishes the human brain is the size of our neocortex, the region related to general intelligence and higher function, which expanded rapidly in recent human evolution. But the neocortex did not replace what was there before. It was layered on top. The brain as a whole is, of course, an integrated device and the different parts of the brain do "talk to" each other. But they do not talk to each other in the way that an engineer might have planned if he was building the human brain from scratch. What we have is an old computer with a new operating system. What are the consequences?

First, brain studies show that it is hard to control our impulses. We like to think that we have unconstrained free will and do things only because we decide to do them. But in many situations we have a hard time achieving this ideal, and in some situations we cannot control our behavior much at all. In general, this is the result of subconscious or preconscious processes that we are not even aware of. We can execute complex behavior without any conscious thought. In certain situations, such automatic reactions can be especially powerful. For example, emergencies and emotions can lead us to do things we would never do if we had time to consider our options and weigh the costs and benefits. Of course, we now know that rational decision-making itself requires emotional input (otherwise we can't decide what to do at all). But, especially in modern settings, emotions can get us into trouble as well.[32]

Second, while control may be hard, it is not impossible. We have seen that some people are better at it than others, and many try and succeed in improving. People can, it seems, even learn to control activity in specific regions of the brain. In one experiment, subjects were shown a graph of the activity levels of their insula (a brain area important for emotional processing), buzzing away on real-time fMRI images. They were then asked to consciously try and alter this level of activity while they were shown either neutral or disturbing images, which triggered emotional responses. After some training, people learned to control their insula activity, reducing their emotional arousal and consequently changing their ratings of the images.[33]

Third, although conflicting mechanisms might imply a poor, inefficient method of making decisions, there are reasons to believe that it may in fact be advantageous. At the intuitive level, being of two minds about something can be akin to arguing out alternative policies in a board meeting. Holding different views generates different kinds of arguments, and the very process of debating them can serve to find and fix errors and gaps in the logic and expand the range and appreciation of possible outcomes. Evolutionary biologist Adi Livnat and computer scientist Nicholas Pippenger found that given physiological constraints, decision-making strategies that include agents in conflict with each other can in fact lead to optimal decisions. Thus even though natural selection may be selecting for a good decision-making strategy overall, that good decision-making strategy may be reached by conflicting internal processes. This suggests the interesting possibility that in principle, our selfish instincts pushing us in one direction and our guardian angels pulling us in another may actually be an effective way of guiding our behavior. Having just one would lead to suboptimal decisions.[34]

Finally, there is evidence that it is precisely the evolutionarily new, neocortical areas of the brain that give rise to conceptions of God. Supernatural agent concepts may therefore be intertwined with the higher-level processing mechanisms that exercise rational influence over lower-level desires. Recent experiments have found, for example, that when subjects pray or are asked about God's involvement in the world and his emotional state, fMRI scans show activity in brain areas critical for social cognition and theory of mind. God concepts appear to rely on, and therefore must have co-evolved with, recently evolved areas of the brain.[35]

In sum, neuroscience shows that there are physiological obstacles to controlling our behavior; while there is variation among individuals, people can learn to improve this control; internal conflicts, though they may seem inefficient, may in fact be adaptive; and finally, supernatural agents are part and parcel of our modern neocortical control room. As Bradley Thayer put it, the "prefrontal cortex is the guardian angel of human behavior." Supernatural punishment may lend a hand in this very department.[36]

How does self-control play out at the psychological level? Psychologists Mike McCullough and Evan Carter have explicitly explored this question in relation to religion. They find that beliefs in supernatural agents play an important role in improving self-control, such that "problems related to waiting, tolerating, and cooperating could be resolved without exclusive reliance on social monitoring and policing, or even expensive institutional monitoring and policing."[37]

McCullough and Carter drew on four established processes that psychologists had previously identified as being required for self-regulation: (1) "clear *goals* that are organized so as to permit effective management of conflict among them"; (2) "sufficient *self-monitoring* and/or self-directed attention so that one can detect discrepancies between one's goals and one's actual behavior"; (3) "sufficient motivation, or *self-regulatory strength*, to change one's behavior when discrepancies are detected"; and (4) "effective mechanisms, or *outputs*, for effecting behavioral change." This provided a framework to examine exactly how religious beliefs might affect each of these four pillars on which self-control is built. The overall conclusion was that religion could be a highly effective mechanism of self-control, precisely because religious beliefs contribute to all of these processes.[38]

Empirical experiments have since been conducted to test this model, with positive results. For example, Carter and colleagues found that "more religious people tended to monitor their standing regarding their goals (self-monitoring) to a greater degree, which in turn related to more self-control. Also, religious people tended to believe that a higher power was watching them, which related to greater self-monitoring, which in turn was related to more self-control." In short, as we might predict, religious beliefs aid self-control. This supports the central idea of the book that religion, compared to alternative methods, is an effective way to moderate selfish behavior. And since selfish behavior can be costly, this was likely to have been favored by natural selection.[39]

As psychologist Mihaly Csikszentmihalyi has suggested, "in groups that prefer peaceful solutions to violent ones, individuals whose brains can control impulses are likely to prosper and reproduce more readily." His point was that evolutionary explanations of morality have tended to focus on the relatively easy question of beliefs and behaviors that are good for the group, but the bigger evolutionary question is why selection would have favored individuals who put the group's interests ahead of their own. What's in it for them? To understand this we have to pay attention to the individual's context within a larger human community over the course of our evolutionary history—our habitat was not only a woodland or savannah, it was a social maze of gossiping humans. In this hazardous environment, anyone lacking impulse control would have run the risk of reputational damage or retaliation by other people. Taming our impulses, however, would certainly have been helped along in the right direction by a fear of supernatural punishment. And as we saw in Chapter 4, the gods can be a far more powerful deterrent than any mere mortal—penetrating surveillance even when we are alone, and punishments worse than death.[40]

STRATEGIC COMPETITION: A GAME THEORETICAL FRAMEWORK

Can we test the idea that there may be adaptive advantages to holding beliefs in supernatural punishment? Let's consider a simple game theoretical framework that pits alternative strategies against each other. This allows us to consider their key differences in terms of costs and benefits to an individual's Darwinian fitness, and identify the broad conditions under which a belief in supernatural punishment may evolve.[41]

THE GOD-FEARING STRATEGY

Even in a world where selfish behavior is costly from an evolutionary perspective, the temptation to cheat remains, along with a susceptibility to sometimes act selfishly even if we did not intend to. My proposal is that something extra—a belief in supernatural punishment—served as a check against social transgressions, helping us avoid the real, worldly costs of retaliation and reputation damage stemming from other human beings. Evidently, these real-world costs were not sufficient on their own. For a deterrent to work against the strong current of selfish behavior, we might have needed something *more* observant, *more* certain, and *more* severe than mere human retribution. Gods, of course, fit the bill. And it may be no coincidence that they are somewhat like us but happen to have *superhuman* powers of surveillance, infallibility, and punishment. God-fearing people may, therefore, have had a selective advantage over nonbelievers, who were more likely to transgress and incur the real-world costs of antisocial behavior.

THE MACHIAVELLIAN STRATEGY

So far we have focused on the *disadvantages* of the emergence of theory of mind and complex language: Selfish actions now bring an increased risk of detection and damage to fitness. However, these cognitive innovations also brought opportunities: to *exploit* them. As Bering and Shackelford argued with blackmail and murder, fitness gains could result from manipulating others' knowledge as well as simply navigating around it. Theory of mind and language clearly gave humans a new source of advantage in effectively gathering, retaining, and regulating the flow of social information that had the potential to impact Darwinian fitness (through whatever means possible, including deception, manipulation, threats, and violence). One may therefore postulate Machiavellian strategies that exploited theory of mind and complex language for personal gain, but that were not God-fearing. Rather than reducing selfishness, these individuals might have taken it to new heights and done better than their gentlemanly peers. As Shakespeare reminded us in *Measure for Measure*, "some rise by sin, and some by virtue fall."[42]

Table 6.1 Three strategies come into competition with the advent
of theory of mind and complex language: Ancestral, Machiavellian, and
God-fearing. Grey shading indicates consequences that act *against* genetic
fitness. Machiavellians outcompete Ancestral individuals, but God-fearers
outcompete Machiavellians as long as the costs of selfishness are high
enough ($pc > m$).

Strategy	Theory of mind and complex language present?	Probability of detection (p)	Cost of exposure (c)	Cost of missed opportunities (m)	Payoff
Ancestral	No	High	Same	None	Lowest
Machiavellian	Yes	High	Same	None	Highest (if $pc < m$)
God-fearing	Yes	Low	Same	Some	Highest (if $pc > m$)

WHICH STRATEGY WINS?

Table 6.1 compares the performance of three strategies: God-fearers and
Machiavellians, following the advent of theory of mind and complex
language, and Ancestral individuals, representing the previous evolutionary state in which people lacked theory of mind and complex language. Machiavellians would clearly outcompete ancestral individuals
because, while everything else is identical between them, Ancestrals do
not have theory of mind and complex language and thus cannot exploit
them for their own gain. More important, however, is that God-fearing
strategists can outcompete Machiavellians. Let's examine how.

Machiavellians and God-fearers differ in just two respects.
Machiavellians enjoy more opportunities for selfish rewards, but have
a higher probability of detection and its associated costs. Therefore
God-fearers can outcompete Machiavellians, *as long as* the total
expected costs of selfishness—namely, the probability of detection
(p) multiplied by the cost of exposure (c)—is greater than the cost of
missed opportunities for selfish rewards (m). This would occur wherever the rewards of selfishness were relatively small compared with its
costs (whether arising from retaliation or reputational damage). Even

a small probability of detection can mean selfishness does not pay if the costs of exposure are high enough. It may not be worth the risk.

The heightened probability and costs of exposure for social transgressions, therefore, may have favored the evolution of traits that suppress, rather than promote, selfish behavior. Indeed, we might expect to see traits that encourage the kind of moralistic behavior that is, after all, empirically common among human societies. My proposal is that *supernatural* punishment was just such a trait. This almighty deterrent helped to steer us away from the real-world punishment that could be so costly to Darwinian fitness.[43]

IS GOD EXCESSIVE? ERROR MANAGEMENT AND THE EVOLUTION OF BELIEFS

One might still wonder why a *supernatural* belief would really be needed to deter selfish behavior. In their review of the role of supernatural punishment in the evolution of religion, biologist Jeff Schloss and philosopher Michael Murray end by asking why, if selfish behavior was so costly, evolution did not favor a simpler solution for suppressing selfishness than the "excessive deterrence of belief in all-knowing, all-powerful agents." I suggest that there are several good reasons why God—or other supernatural agents—actually represent the best possible deterrent against selfishness.[44]

God may work precisely *because* he is excessive. A belief in supernatural punishment may be a good method of deterring people from the real-world costs of selfish actions because of an asymmetry in the possible errors one may make. Beliefs can be wrong in two ways: believing X is true when it is false (a false positive error), and believing X is not true when it is (a false negative error). "Error management theory" (EMT) suggests that wherever the costs of false positive and false negative errors are asymmetric, we should expect natural selection to favor some method of avoiding the more costly error. For example, we often think a stick is a snake (which is harmless), but we do not tend to think snakes are sticks (which is potentially deadly). This is obviously a highly adaptive bias. Similarly, when we set the sensitivity of a smoke alarm, it is better to err on the side of caution because the costs of a false alarm are negligible (however annoying it may be at the time), whereas the costs of being burned to death in a fire are great indeed. One should thus expect the smoke alarm to go off a bit too often. Whenever the true probability of some event—such as the detection of snakes, fires, or selfish behavior—is uncertain, a biased decision rule can be better than an unbiased one if we want to avoid the more dangerous error. To be clear, the reason why a bias is best is because an *unbiased* decision rule centered on the true probability of the event will make both false positive and false negative errors—it will generate some false alarms, and it will fail to go off in some real fires. By contrast, a *biased* decision rule will err on the side of caution—it will have a lot of false alarms, but more importantly it will rarely fail to go off in a real fire. This logic of managing errors is thought to underlie a range of human cognitive biases, including overconfidence, assuming the worst in others, and differing perceptions of sexual interest among men and women.[45]

Plugging this logic back into the context of religion, if we aim to assess the true probability of detection for cheating, then half the time we will overestimate the probability and get away with it, and half the time we will underestimate it and get caught. That might seem a reasonable, fifty-fifty strategy. But the kicker is if these two mistakes entail different costs. If false negative errors (assuming stealth and getting caught) are more costly than false positive errors (assuming detection and missing a reward), then *only exaggerated* estimates of the probability of detection—such as a belief that supernatural agents are observing your behavior all the time—will help you to avoid the worst of the two errors. As long as there is some uncertainty about the true probability of detection, simply improving accuracy will not help. The best solution to avoiding detection, therefore, is a mechanism that overestimates the true probability that detection will occur—*exaggerated* estimates outperform *accurate* estimates, because the latter will engender more mistakes.[46]

This means that it is not just an asymmetry of costs and benefits that may be decisive (God-fearers doing better than Machiavellians), but a bias to *overestimate* what those costs and benefits are (God-*fearing* itself is favored by natural selection). Given the dangers of our social mine-field, an exaggerated belief that one is constantly being watched and judged by supernatural agents—the fabled hyperactive agency detector device (HADD)—may be an especially effective guard against careless selfishness. Schloss and Murray's concern that God is a "seemingly excessive" deterrent is prescient—the threat of punishment may *need* to be excessive, such as a belief in an omniscient and omnipotent God, because this is the most effective way to avoid dangerous mistakes.[47]

God is also a good deterrent because supernatural agent concepts are easily accessible and fast. The cognitive science of religion literature sug-gests that beliefs in supernatural agency and supernatural consequences are the cognitive default, and are more easily accessed than many other concepts. In particular, it appears that negative, incriminating informa-tion is particularly salient and rapidly available to people in tests that are designed to measure people's subconscious, implicit associations. Other forms of deterrent, if they are (or once were) rivals, might therefore have been trumped simply by the ease and speed of supernatural cognition.[48]

The idea that a cognitive bias or cognitive illusion may be a good way to achieve adaptive behavior is not so bizarre as it may sound. In fact it is turning out to be a commonly recognized phenomenon in psychology. There is a huge list of human cognitive and motivational biases, and many of them are thought not to be malfunctions or con-straints of the human brain, but rather adaptive heuristics that help us solve important problems. In a complex world with a mesmerizing barrage of information and the need to navigate through it efficiently, the brain has developed numerous shortcuts that serve to quickly and effectively direct our behavior. In the previous chapter we saw sev-eral examples of apparently irrational biases that underlie supersti-tious behavior in all of us. But even superstitious behavior can have advantages if it somehow reduces anxiety, helps us attempt otherwise daunting tasks, or improves performance. As has been demonstrated in laboratory experiments, superstitious beliefs can actually increase success at solving problems. In the real world, too, there is evidence that superstitious beliefs can help us out. One study, for example,

found that higher performing basketball teams, and higher performing players within them, showed more superstitious behaviors. We also
saw that evolutionary biologists Foster and Kokko suggest superstitious behaviors are common among animals, and this is no accident:
they were favored by natural selection because they increase fitness. As
David Sloan Wilson has argued, "the unpredictability and unknown
nature of our environment may mean that factual knowledge isn't as
useful as the behaviors we have evolved to deal with this world." For
this reason, to understand adaptive behavior we should not look for
accuracy or rationality, but rather for what works. "Adaptation is the
gold standard against which reality must be judged," Wilson notes. By
this standard, God is a great solution.[49]

CONCLUSIONS

In the court of fourth-century B.C. King Dionysius II of Syracuse, Sicily,
there was a man called Damocles who was in awe of his king. The
king lived in great luxury, reigned with power and authority, and was
surrounded by great men. Damocles counted the king a lucky man.
Intrigued, Dionysius suggested they switch places so that Damocles
might see what being king was really like. Once seated in the throne,
Damocles surveyed his position of greatness and was distraught to find
a heavy sword hanging directly above him, dangling by a single horse's
hair. The sword threatened to drop and kill Damocles at any moment.
Horrified, he pleaded to switch back again. He no longer wanted to be
so lucky as the king. The Sword of Damocles, as Dionysius's lesson told,
was the ever-present threat of lethal danger that a king must endure
when all eyes are upon him. No matter what his luxuries and power, or
indeed precisely because of them, he feared every moment for his life.

  We do not need to be powerful kings to live under the burden of scrutiny and threat. Supernatural punishment is like the Sword of Damocles.
If there is a God, then all of us are under watchful surveillance and the
possibility of supernatural punishment. We go through life with freedom and tempting opportunities, but if ever we make a misstep, there

is a sword waiting to strike us down. In fact the situation is worse than for Dionysius, because there are two swords raised above us. One is the relatively blunt instrument of real-world consequences—the retaliation or injury to our reputation we may receive from our fellow man. This sword is dangerous, and damaging to Darwinian fitness, but it is erratic and not usually lethal. The other is the weighty and sharp instrument of divine punishment, in this life or the afterlife, that we may receive from the gods. This sword is truly terrifying, with a blade that is infallible and can deal a fate even worse than death. For a believer, life is doubly dangerous, and the sharp, heavy sword of God is the one that elicits the greatest fear and is the greatest deterrent. But this is exactly where natural selection may have extracted real-world benefits from supernatural beliefs. The blunt sword of our fellow man is real, and the sharp sword of God may not be, yet a fear of the sharp sword can save us from the blows of the real one. A belief in the wrath of God steers us away from risking the wrath of man.

Because egoistic behavior is deeply ingrained and hard to control, we may need encouragement to stay on the rails. Rational restraint is often too weak, too slow, or simply unavailable. This is where God can help us, whether he is real or not. People who fear supernatural punishment—those living with a supernatural Sword of Damocles above their heads—are likely to be more cautious in their behavior, and as a result may incur lower real-world costs.

While writing this book I asked a colleague trained in theology and religious studies whether there was a standard work in the field on the role of supernatural reward and punishment that I might have missed. He paused and thought for a minute, then said, "no I don't think there is anything like that. It sounds more like economics." This was a telling observation because it is precisely the economics of costs and benefits that underlies an evolutionary perspective on religion. What evolutionary biology does is scrutinize an organism's traits from the perspective of the cold economic accounting machine of natural selection, and explores how investments in these traits contribute to dividends or deficits in reproductive success. Costly investments that fail to bring returns cannot survive in a competitive marketplace. As Richard

Dawkins notes, evolution quashes any tiny extravagance because "natural selection abhors waste." It is economic gains and losses, therefore, that fundamentally underlie the evolution of our physiology, brains, and behavior. If a trait doesn't pay, it usually falls by the evolutionary wayside. Indeed, paleontologist Geerat Vermeij called his treatise on the 3.5 billion year history of the evolution of life on Earth *Nature: An Economic History*. The simplicity of natural selection is what unnerves many people about it. But that is also its power. It offers a fresh insight on all human affairs because it forces us to think hard about how any given behavior would have had positive or negative consequences for fitness in our evolutionary past. Without thinking that through, we cannot develop a clear picture of whether something we observe today is an adaptive trait favored and shaped by evolution, or a mere byproduct of other traits or a recent innovation of culture. We all know that supernatural reward and punishment plays some role in the everyday lives of billions of believers around the world today. The bigger question is what effects these beliefs may have had in our evolutionary history. Obtaining and interpreting data on our earliest ancestors, and on contemporary hunter-gatherer societies as analogues of what life was like for our Pleistocene ancestors, is not easy. Nevertheless, I suggest that a belief in supernatural punishment helped, not hurt, the economic bottom line of Darwinian fitness, and insodoing pushed human beings along a path to lower costs of selfishness for themselves and greater benefits of mutual cooperation with each other.[50]

Einstein, in reflecting on religion, noted a simple but fundamental feature of life that pervades all else: "Strange is our situation here on Earth. Each of us comes for a short visit, not knowing why, yet sometimes seeming to divine a purpose. From the standpoint of daily life, however, there is one thing we do know: that man is here for the sake of other men—above all for those upon whose smiles and well-being our own happiness depends." Just as Seneca noted in the opening quote, doing good to others can be good for ourselves. And this, as we shall now see, watered the seeds of civilization.

CHAPTER 7

NATIONS UNDER GOD

Those who deny the existence of the Deity are not to be tolerated at all. Promises,
covenants and oaths, which are the bonds of human society, can have no
hold upon or sanctity for an atheist; for the taking away of God,
even only in thought, dissolves all.

—*John Locke*[1]

At the beginning of the previous chapter, we learned about the decline of the Shakers, whose belief in celibacy is leading them to extinction. But other religious groups tell exactly the opposite story—having beliefs that helped them survive. The Puritans, persecuted in Europe, might have disappeared altogether had they not braved the Atlantic Ocean to establish a new home in an unknown wilderness now called the United States of America. It was a highly risky venture. But the motivation to pursue their freedom against the odds, and to stick together through tough times, was deeply rooted in their religious beliefs and their faith in God. With the benefit of hindsight the achievements of the pilgrims can seem a historical inevitability, but at the time there was no great reason to expect success. It was, as Nathaniel Philbrick put it, "a stunningly audacious proposition." Apart from Jamestown, "all other attempts to establish a permanent English settlement on the North American continent had so far failed," and even the settlers in Jamestown had buried more of their fellows than survived in its first years. The Puritans had even fewer resources than most of those who had come before, and to make matters worse there had been an upsurge of lethal Indian attacks against settlements up and down the Atlantic seaboard. Yet they harbored a belief that they could prevail. As two of their leaders wrote, "it is not with us as with other men ... whom small things can discourage, or small discontentments cause to wish themselves home again."[2]

Their exploits and achievements were truly remarkable, and were no doubt aided by their strong beliefs in a greater mission and their religious duty to members of the faith—helping each other to survive the voyage, find food, build the settlement, nurse the sick back to health, and foster each others' children when people began to die. William Bradford, first governor of Plymouth Colony and author of the famous memoir *Of Plymouth Plantation*, was himself struck by the selfless deeds shown by the Puritans in the deadly first winter: "Whilst they had health, yea or any strength continuing, they were not wanting to any that had need of them." Among the Puritans were also the crew of the Mayflower and the so-called Strangers, recruited by financiers of the venture to help establish the colony. The contrast sometimes showed. As the Mayflower's boatswain lay dying, attended to by the very Puritans he had treated dreadfully from the start of the voyage, he cried out: "Oh, you, I now see, show your love like Christians indeed one to another, but we let one another die like dogs."[3]

Even once the colony was established, the struggle was far from over. Their tenuous grip on the New World made them especially keen to transmit their beliefs to subsequent generations. Only they would be able to ensure the future of Puritan society. And they were not shy about invoking the idea of supernatural punishment to help this effort. As Melvin Konner noted, "the severity with which the Puritans imbued their children with the fear of God was in part a response to an ever-present threat of mortality that made heaven and hell seem only a breath away. Families were constantly adjusting to deaths, especially the deaths of children." In their first winter, half of the colony had died. Konner points out that mortality may have been just as high in other societies of the era that "did not burden children with guilt and fear." However, "given the Puritan's beliefs, they had to try and ensure their children's salvation; even their harshest punishments would be kind compared to damnation." Remarkable as such beliefs appear today, they may have been instrumental in fostering the cooperation necessary to survive, and later grow, against the odds (Figure 7.1). As Philbrick explains, "a Puritan believed that everything happened for

Figure 7.1 "The First Thanksgiving," Jennie Augusta Brownscombe (1915). Would the pilgrims have thrived or even survived in America, against the odds, had it not been for their strong faith generating a highly disciplined, determined, and cooperative society? Courtesy of Pilgrim Hall Museum, Plymouth, Massachusetts.

a reason. Whether it was the salvation of John Howland [fished out of the Atlantic in a storm] or the death of a profane young sailor [who was the first to die], it occurred because God had made it so." But the plight of the Puritans was not about *individuals* surviving or doing better than others. An individual could never have survived the crossing, or the winter to come. The Puritans' story is not so much about individuals and their religious beliefs, as about a religious society.[4]

In the previous chapter we examined what religion does for individuals. For cognitively sophisticated humans, pure self-interest brings social costs that religion may have helped to avoid. There were, therefore, likely to have been evolutionary advantages to individuals whose selfish behavior was deterred or decreased by a fear of supernatural punishment. However, the greatest impact of our payback mindset is on society as a whole. All those individuals with selfishness ratcheted down by natural selection means there are gains for the group as well. If everyone is less selfish and more cooperative, the collective will benefit as well as the individuals within it—and the gains may be greater than the sum of the parts.

This development opened up whole new worlds of possibilities for human endeavor. Societies could become larger, since cooperation could be more effectively sustained even in large groups. They could develop more complex divisions of labor, since people could expect others to share resources, trade reliably, and be more likely to honor their commitments. And they could achieve much greater feats of collective action, allowing such marvels as roads, buildings, standing armies, and ultimately empires spanning thousands of miles and multiple ethnic groups. Other factors such as agriculture, technology, and writing were also clearly key to the social revolution of the last 10,000 years or so. But beliefs in supernatural, punishing agents may have played a formative role in the origins and growth of complex human societies. Without them, cooperation in large groups would have been that much harder to achieve. It might even have been impossible. Ara Norenzayan's research on the development of human civilizations since the emergence of agriculture argues that powerful "big gods" provided an essential mechanism for large-scale cooperation among strangers and in which, he stressed, "the action is in the fear of supernatural punishment."[5]

RELIGION AND SOCIETY

The idea that religion is essential to society goes back a long way. The word "religion" itself is usually a descriptor of a cultural phenomenon, not an individual one, and indeed has its root in the latin "religare," meaning to bind together. Until now our discussion of beliefs and consequences—the fear of supernatural punishment, and its effects on Darwinian fitness—has focused on individuals, individual psychology, and individual selection. But those individuals are not living in a vacuum. Individual beliefs take on even greater significance and power when they are plugged into a group setting. And they seem to systematically help, rather than hurt, the group.

English philosopher John Locke, who played an instrumental role in the Enlightenment, worried that without religion society would cease to function altogether. Other influential thinkers such as German philosopher Immanuel Kant noticed that whatever their own views, people were reliant on the effects of religious belief *on others* to make society work. The moral behavior that religion could inspire was important for our own social navigation, but it was possibly even more important for establishing trust with other people—if *they* did not believe in God, then what value could their "promises, convenants, and oaths," as Locke put it, have?[6]

This question took on a special relevance in the nineteenth and twentieth centuries as anthropologists began making the first detailed studies of cultures outside the western world. Most notably, French sociologist Emile Durkheim, who was to become one of the fathers of the discipline, immersed himself in ethnographies of the aboriginal peoples of Australia. From this, Durkheim developed the foundational idea that religion is a kind of glue that binds society together. Or, as he famously put it as a definition of religion, "a religion is a unified system of beliefs and practices relative to sacred things, that is to say, things set apart and forbidden—beliefs and practices which unite in one single moral community called a Church, all those who adhere to them." Moreover, he did not think that religion developed by accident. Religion provided a practical means by which people could effectively organize communal activities, which helped them in numerous ways to subsist and survive. Though their beliefs may have been supernatural, they had what David Sloan Wilson calls "secular utility"—tangible material payoffs in the real world.[7]

This kind of functionalist account—trait X is there in order to solve problem Y—fell out of favor later in the twentieth century. Not everything has to be *for* something, critics pointed out. Many social anthropologists preferred to document cultural variation without searching for causal patterns or purpose. Certainly, some features of culture seem to just float around without having any strong positive or negative effect on individual or group success. They may come and go, but they don't do anything much in terms of contributing to a society's success or failure. However, other cultural traits do very much seem to have important consequences for group life, and if so their patterns and purpose become crucial to our understanding of society. Function can matter not only to group success, but also to our success in understanding those groups.

In recent years functionalist accounts have been powerfully revived with the new tools of modern evolutionary biology. Rather than just expressing an opinion on whether cultural traits may be good for people or not, or avoiding the topic altogether, today we can go out and *test* the effects of cultural variations on survival and reproduction (or on behaviors that are likely, or were likely in the past, to have contributed to survival and reproduction). Behavioral ecologists and evolutionary anthropologists have begun systematically collecting cross-cultural data to examine exactly this kind of question. Do religious beliefs and behaviors help us or not? It would be no surprise to Durkheim that many of these studies show that religion does indeed help people—in spite of, or perhaps precisely *because* of, all its costs and peculiarities. People invest time, effort, and resources in religious activity because it generates great dividends. Religion can help individuals, but its positive effects are powerfully magnified at the larger level of society. Kant and Durkheim appear to have presaged what evolutionary theory would later corroborate: Religion serves to bind individuals together, and this opens the way for cooperation and collective action to reach new heights.[8]

THE SIMPLICITY OF PUNISHMENT AND THE COMPLEXITY OF RELIGION

Religions are extremely intricate cultural systems. Indeed, they may be examples of complex adaptive systems—social or physical phenomena that contain not only numerous elements, but numerous interconnections between these elements that are themselves integral to the whole. No single psychological propensity toward religious belief or behavior can explain *religion* as a whole—the complex socio-cultural

phenomenon of beliefs, narratives, myths, obligations, rituals, organizations, structures, history, and so on, let alone their interactions.
These are clearly more than the sum of individual cognitive mechanisms or behaviors. As even Richard Dawkins notes, "religion is a
large phenomenon and it needs a large theory to explain it." Amidst
the formidable complexity, therefore, how much can the crude instrument of supernatural punishment really do to help explain the origins
and evolution of big, complex religions?[9]

While supernatural punishment can't explain all of religion, it does
appear to be a master lever that exerts a powerful effect on other elements in the system. It also interacts powerfully with other phenomena,
such as leadership, to leverage large social effects that would not otherwise be possible. And it serves to add weight to other mechanisms that
are converging means to the same end, such as supernatural rewards
and secular punishment. Religion is complex, but this complexity means
supernatural punishment can have multiple effects and reinforcements
as well as complications and resistance. This chapter examines ways in
which supernatural punishment promotes cooperation in society not
just in spite of the complexity of religion, but often because of it.

CROSS-CULTURAL EVIDENCE THAT RELIGION PROMOTES COOPERATION

The claim that religion helps society is often heard. But what actual
evidence do we have to support it? Does religion really promote higher
levels of cooperation? And what evidence is there that a driving factor,
among the many features of religion, is the prospect of supernatural
punishment?

The most immediate evidence on these questions comes from the
major world religions' own doctrines. The origins, development, and
contemporary concerns of religions such as Judaism, Christianity,
Islam, Hinduism, and Buddhism are candid about promoting things
that are good for society as a whole—unity, community, self-restraint,
sharing, generosity, helping others, and so on. And they are often

equally candid about the consequences of failing to follow such pre-scriptions, or indeed rejecting the faith altogether. There is, of course, a long-standing debate over *who* exactly benefits from injunctions to cooperate. For example, in Christianity, various lines of evidence sug-gest that even Jesus's call to "love thy neighbor" did not mean, as often assumed, that "neighbor" meant everyone else in humanity. Rather, it seems to have been a call to direct help and generosity to other people *within the in-group*. But however near or far such calls are supposed to extend, most scholars agree that religion is intended to help at least those in your immediate community, whether or not it also helps oth-ers beyond it.[10]

Is this direct instruction to cooperate just a feature of modern soci-eties and the major world religions? What about indigenous societies and ancient civilizations? And in any case, however well-intentioned religion may be in *advocating* social cooperation, does it actually suc-ceed? It turns out that wherever one looks, one finds religious beliefs and behaviors are primarily directed toward helping out society and the people within it. And it works.

Let us begin with indigenous societies. Earlier we looked at religions in sub-Saharan Africa, and here the message rings loud and clear. Hadnes and Schumacher found traditional beliefs "mostly involve supernatural punishment for any behavior that contradicts the moral code of the community. Such a moral code contains customs, regula-tions, and taboos to ensure the cohesion of the society and the con-tinuation of the ancestral tradition." Beliefs are inextricably linked to the social good: "Moral codes define the conduct of each individual within the community where the community includes the living and the departed (the ancestors). Probably, the most prominent social norm is the subordination of the individual to the community."[11]

The powerful linkage of religion and social norms in Africa has been an important issue for western efforts at economic assistance, business development, and aid. Professionals in developing countries have often seen traditional beliefs and practices as obstacles, prevent-ing the adoption of "better," more efficient, western ways of think-ing and working. However, these newly imposed methods often turn out to be counterproductive. Traditional beliefs form a crucial social fabric into which anything else must fit: "While most development researchers and practitioners claim that traditional African beliefs obstruct economic growth, scholars of African studies and anthropology suggest that they can equally help to overcome social-dilemma situations when formal institutions regulating economic transactions are missing." In short, where the tools of modern governance and organization may seem lacking, religions—and especially supernatural punishment—turn out to be the very mechanisms that help solve the problems of cooperation. And the pattern extends across the globe.[12]

Elinor Ostrom's studies of how indigenous people around the world solve collective action problems and foster the sustainable use of "common pool resources"—resources that must be preserved and shared by groups of people—is especially enlightening. In a world of climate change, dwindling resources, and overexploitation of fisheries, forests, and water, Ostrom's work has gained increasing prominence and importance. The question everyone wants to know the answer to is, how do they do it? Over many years of fieldwork observing these societies in action, Ostrom identified a set of eight common factors or "design principles" that are critical for the effectiveness of such cooperation:[13]

1. Clearly defined boundaries of the resource and the people with the right to use it.
2. Mechanisms ensuring the equity of effort and benefits for members.
3. Collective-choice arrangements so that decisions are made by consensus or another system agreed as fair.
4. Monitoring of agreed behavior.
5. Graduated sanctions in response to violations.
6. Conflict-resolution mechanisms that are fast and fair.
7. Local autonomy for a group to manage its own resources.
8. Replication of the above principles when part of a larger hierarchy.

Ostrom convincingly demonstrated that local ways of doing things can be remarkably effective. However, while considerable research has been conducted on the problem of common pool resources, there had until recently been very little work on the role that *religion* might play.

New work has looked specifically at the role of religion in these common pool resource problems. And it turns out that supernatural punishment is key.

In a study of forty-eight indigenous societies engaged in the management of common pool resources—from all continents and a variety of habitats ranging from forests to rivers to lakes—a research team from Binghamton University in New York State found that *religious* mechanisms of monitoring and enforcement were crucial to all of Ostrom's eight design features. Not only were they even more common and important than secular methods, but a dominant feature was an explicit regime of *supernatural* surveillance and sanctions in which: "Supernatural agents are perceived to play a large role in monitoring and punishing deviant behavior." People in these communities harbored a genuine concern that spirits or other supernatural agents would punish violations of group norms on how resources were to be shared.[14]

These studies suggest that religion is not just present in social cooperation, but primary and crucial. It even trumps secular versions of monitoring and punishment, where those are present too. Most notably, supernatural punishment plays a critical role in preserving precious resources that the community as a whole must share if the resource, and the society depending on it, is to survive. Whatever religion may do for individuals, it can be an essential means for society to function and flourish.

My own empirical work comparing 186 preindustrial cultures around the world also found support for a link between beliefs about supernatural agents and cooperation. The overall result was that irrespective of the *type* of religion or *region* of the globe, moralizing gods were more frequent among societies that were larger, centrally sanctioned, policed, use and loan money, and pay taxes. The popular theory among many cognitive scientists of religion, that religion is an arbitrary byproduct of big brains or cultural drift, does not predict any relationship between cooperation and moralizing gods. But in fact, societies with more moralizing gods were more cooperative, precisely as the supernatural punishment hypothesis would predict.

These are only correlations, of course, so we don't know if moralizing gods *caused* more cooperation (or vice versa), but it is consistent with the theory that the two should be associated.[15]

These cross-cultural studies lend empirical weight to the idea that supernatural agents promote cooperation, one critical outcome of which is allowing societies to become large. However, this leads to a puzzle: How do we explain the statistical outliers—societies that became large, and thus presumably highly cooperative, even though they *lacked* moralizing gods? How is social cooperation promoted and internal conflict suppressed in those societies? Exploring these exceptions to the rule offers an alternative way of examining the theory.

The most extreme outliers in the data from my study were the Roman and Babylonian civilizations. Both were huge societies. And both had gods, of course, but these supernatural agents have been coded by anthropologists as much less moralizing than the gods of other societies. As we saw in Chapter 3 this does not mean that belief in supernatural punishment was absent, but rather that compared to other societies gods were less tightly linked to moral behavior. The Romans, for example, consulted philosophers to determine what was morally right or wrong, not gods or religious leaders. Their concerns about supernatural punishment, at least in an afterlife, were also tempered. Tombstones were sometimes inscribed with "nf f ns nc" (*non fui, fui, non sum, non curo*—I didn't exist, I did exist, I don't exist, I have no cares). Without moralizing gods, alternative mechanisms are more likely to have been needed to sustain cooperation. And this is precisely why the Romans are so interesting, because they were characterized by, indeed originated, remarkable and world-changing notions of honor, duty, citizenship, and intricate systems of governance that served to promote cooperation. The Babylonian civilization solved the problem in a different way. It reached its zenith under the Amorite king Hammurabi, who is widely known for his Law Code, an extraordinary set of detailed social rules that were accompanied by brutal punishments. In such a system of harsh discipline, it seems, Hammurabi did not need gods to do his dirty work for him.[16]

These remarkable examples complement the broader results of the cross-cultural studies. Though outliers, they appear to be exceptions that prove the rule: Without moralizing gods (or at least where gods were less moralizing), societies needed very powerful alternative mechanisms of social control if they were to function and flourish.

Empirical evidence from more recent historical communities also shows a positive effect of religious beliefs for society. Anthropologists Richard Sosis and Eric Bressler compared the long-term survival of dozens of nineteenth-century utopian communes in the United States. These were communities of people who, for shared ideological reasons, isolated themselves from the rest of American society to live and work together in their own way of life, such as Hutterite, Owenite, anarchist, and socialist groups. Some were religious and others were secular. Sosis and Bressler found that religious communes were three times more likely to survive than secular communes, and this appeared to result from higher levels of cooperation. The number of costly investments and sacrifices made by religious communities predicted how long they survived—the more people had to invest and give up, the more likely the commune was to survive. But strikingly, the number of costly investments and sacrifices did *not* explain how long *secular* communes survived. Simply demanding or banning things did not achieve the necessary levels of cooperation; such injunctions appeared to need an underlying religious belief system to support them effectively. Sosis and another colleague, economist Bradley Ruffle, found a similar effect when comparing secular and religious Kibbutz communities in Israel. As has been widely documented, religious kibbutzim are doing very well, while secular ones have suffered serious economic decline. The missing factor appears to be religion, which somehow manages to boost cooperation and boast greater success.[17]

A number of other studies have looked at the positive or negative effects of religion among contemporary societies. Overall, countries that are more religious tend to have *lower* economic performance, but this correlation is driven by the prosperous West which, perhaps for independent reasons, is also more secular. When societies within similar development brackets are compared, the relationship is the other way around: Higher levels of religiosity are associated with higher economic performance. For example, economists Robert Barro and Rachel McCleary compared a large sample of countries and found that beliefs in heaven and hell were greater among countries that had stronger economic growth. And it really did seem to be beliefs rather than practices that drove the relationship, because mere attendance at religious services actually worked in the opposite direction. As we saw in Chapter 3, another large cross-cultural study using the World Values Survey has found that among a huge sample of over 100,000 people in dozens of countries, people who believe in an afterlife express less tolerance of moral transgressions. Such studies are hardly direct evidence that belief in supernatural punishment *causes* greater cooperation, but given the sample sizes and the global spread of the data, it is a significant finding that operates in exactly the direction we would expect given the argument of the book. Evidence also suggests that religious beliefs in general are stronger in societies that suffer from greater levels of insecurity—situations in which cooperation and internal order may be, or are perceived to be, especially important.[18]

Psychologists Azim Shariff and Mijke Rhemtulla found more direct evidence of a link between supernatural punishment beliefs and cooperative behavior. Comparing different countries, they found that the more a nation's population believes in hell (on average), the lower the crime rates. There are clearly many other potential factors at work here, but the correlation remained significant even after controlling for several other well-established variables. Certainly, the relationship is again pointing in the direction we would expect if belief in supernatural punishment has a positive effect on society.[19]

We can see for ourselves many examples of how religion benefits communities and societies—even if it may sometimes also contribute to conflict within or between them. Christian churches support local communities and care for the destitute and sick. Islam fosters the giving of alms to the poor and connects followers together in a wider community. Judaism unites adherents from multiple ethnic communities and diasporas around the world. Buddhism teaches generosity and good deeds and the relegation of individual desires. Rather than the question of whether religion can be good for society, the plentiful social benefits of religion raise the question of how well societies would survive without it.

There are interesting natural experiments going on today in which religion is declining, such as in Western Europe and especially Scandinavia. Do these examples undermine or support the argument? First, although these societies seem to thrive without high levels of religious adherence, religious *beliefs* have often not dropped as low as church *attendance* might suggest. In many ostensibly secular nations, what we would call religious beliefs and practices are alive and well. Second, in these societies, institutional mechanisms and community initiatives are replacing the roles that religions and local churches played. With religion in decline, government is stepping in where God has withdrawn. This suggests that religion *was* important, and left a yawning gap when it declined. If you take out God, you have to put something else back in. But the jury is out on whether these alternatives are better than God, about the same, or worse. Secular methods of maintaining social cooperation may work fine, but will they ever match the effectiveness of an all-knowing and all-powerful God?[20]

PUNISHMENT IN THE NAME OF RELIGION

The focus of this book has been supernatural punishment, and how it steers people's behavior. However, sometimes compliance with religious rules, taboos, and commandments is enforced not by God, but by people. Whether or not breaking sacred rules is punishable by supernatural agents, they may also be punished by society *in the name of* the religion. In many preindustrial societies, it is human beings—spiritual headmen, priests, chiefs, kings, or their assistants—who monitor and sanction people for flouting religious rules. As psychologist Matt Rossano found, "often, the shaman finds that violations of taboo have aroused supernatural anger. By binding supernatural authority to the punishments incurred when taboos are broken, the shaman reinforces group norms." In our evolutionary past, therefore, supernatural beliefs may not only have been important as a psychological deterrent, but as a justification for secular deterrents.[21]

The phenomenon continues in some societies today. Saudi Arabia has approximately 5,000 agents of the Commission for the Promotion of Virtue and Prevention of Vice—the *hay'ah* or *mutaween* (meaning literally, "volunteers"). The agents ensure that shops close for each of the five daily prayers, that people dress appropriately, and that men and women are strictly separated in public. They also "prevent sorcery and round up bootleggers and drug dealers."[22]

In recent years there has been a rise of complaints about the agents' "excessive zeal." People had died in their custody, and for the first time ever, someone had filed a lawsuit against them. In 2002, the *mutaween* attracted especially fierce criticism after claims that it blocked exits to a girls school in Mecca during a lethal fire, because the students were not dressed appropriately to go outside. Several died and dozens were injured. Only at this point did new rules come into force. The agents now had to wear badges, stop carrying sticks, hand over suspects to the regular police, and attend training sessions to improve public relations.

Interestingly, however, as well as criticism, the agents receive a good deal of help. They "can count on many unofficial helpers to tip them off. The regular police generally sympathize with their fellow law-enforcers and can be conveniently slow to arrive when called to pick up suspects. *Mutaween* supporters have fended off harsher legislation that would have required agents to wear uniforms, as well as calls for their duties to be defined within specific limits." The Saudi state identifies the Qur'an as its constitution, and the "mission of promoting virtue and preventing vice ... [is] a scriptural Islamic injunction." Shari'a law itself is, of course, the enforcement of religious prescriptions and proscriptions.[23]

The Saudi religious police may be an extreme example, but the enforcement of religious laws by society is found in many other cultures and religions. In the case of Christianity, according to the *Wycliffe Dictionary of Theology*, "throughout the bible it is insisted that sin is to be punished. In an ultimate sense God will see that this is done, but temporarily the obligation is laid upon those in authority to see that wrongdoers are punished."[24]

Until recent times, Christianity was used as an explicit justification and basis for the law. Specific religious offenses, such as blasphemy, could be listed as crimes and punished accordingly. Indeed, religious crimes were sometimes afforded their own special punishments. This history has been gleefully highlighted by the New Atheists such as Sam Harris, who noted: "The medieval church was quick to observe that the Good Book was good enough to suggest a variety of means for eradicating heresy, ranging from a communal volley of stones to cremation while alive. A literal reading of the Old Testament not only permits but *requires* heretics to be put to death." And, Harris claims, "it was never difficult to find a mob willing to perform this holy office, and to do so purely on the authority of the Church." Lest this be taken as an aberration, Harris adds that "showing a genius for totalitarianism that few mortals have ever fully implemented, the author of this document [Deuteronomy 17: 12–13] demands that anyone too squeamish to take part in such religious killing must be killed as well." Harsh as this may seem and as unlikely as it is that it was regularly or reliably carried out, it is a striking attempt to address the problem of second-order free-riders that we met in our discussions of game theory earlier—people who may be willing to cooperate themselves but are unwilling to punish noncooperators. Harris goes on to note that religion served as a justification for punishment even among theologians such as St. Augustine, "who reasoned that if torture was appropriate for those who broke the laws of men, it was even more fitting for those who broke the laws of God." However stunning it may seem from a modern perspective, the intended (if not actual) effects on society are plain: "There really seems to be very little to perplex us here. Burning people who are destined to burn for all time seems a small price to pay to protect the people you love from the same fate."[25]

Things are very different today, of course, but that is a recent development, coinciding with the advent of institutions of law, police, courts, and jails. These have only appeared in the last few hundred years, and have only become secular and transparent in the last few decades (or still not at all in many countries). Even then, these institutions are influenced a great deal by their underlying culture and the religious traditions therein.

Though punishment in the name of religion may be rare in western societies today, many vestiges of this history remain in the way that originally religious moral standards have been codified into law and cultural norms. And there may be deeper psychological origins as well. The Princeton physicist Freeman Dyson, who grew up in Britain, suggests that people have a bias toward advocating, participating in, and celebrating punishment in particular. While Americans are out trick or treating in the fall, English children are out a few nights later burning effigies of Guy Fawkes, the instigator of the gunpowder plot who was caught red-handed beneath the Houses of Parliament in London:

> For a child in England, there are two special days in the year, Christmas and Guy Fawkes. Christmas is the festival of love and forgiveness. Guy Fawkes is the festival of hate and punishment. Guy Fawkes was the notorious traitor who tried to blow up the King and Parliament with gunpowder in 1605. He was gruesomely tortured before he was burnt. Children celebrate his demise with big bonfires and fireworks. They look forward to Guy Fawkes more than to Christmas. Christmas is boring but Guy Fawkes is fun. Humans are born with genes that reward us with intense pleasure when we punish traitors. Punishing traitors is the group's way of enforcing cooperation. We evolved cooperation by evolving a congenital delight in punishing sinners.[26]

We might debate the relative attraction of Christmas and Guy Fawkes night (a lot of kids prefer presents to the charcoaled remains of a straw man), and whether we are really gaining pleasure in his eternal punishment (any excuse for a bonfire and fireworks), but what certainly rings true is that there is something powerful and scary about grassroots punishment, when a whole village comes out bent on retribution. If the offense is against religious ideals or taboos that are held sacred, the venom can be especially strong—and we have plenty of examples of such religious vigilantism today from Pakistan, Nigeria, the Middle East, and elsewhere.

Finally, religion can be used to justify some kinds of punishment even outside the law. If bad things are done to bad people, so the logic goes, then it may be forgiven. A recent study of beliefs among hard-core street offenders notes that despite their criminal activity, "many such offenders also hold strong religious convictions, including those related to the punitive afterlife consequences of offending." What was striking, however, was how consistently religious beliefs themselves were interpreted to rationalize or even justify crimes. As one twenty-five-year-old drug dealer explained: "If I go rob a dope dealer or a molester or something, then it don't count against me because it's like I'm giving punishment to them for Jesus. That's God's will." And that brings us to the topic of crime.[27]

CRIME AND PUNISHMENT: THE HELLFIRE HYPOTHESIS

Many people believe that without religion, society will be unable to function and fall into anarchy and crime. This simplistic argument has been debunked by many, not least Richard Dawkins. Yet just because religion isn't necessary for good behavior, does not mean that it doesn't help. Supernatural punishment can supplement or enhance whatever deterrence there is from secular punishment.[28]

So are criminals in general less religious than law-abiding citizens? This is dangerous territory, because whatever the relative rates of religious faith, it may be a range of interconnected socioeconomic factors that are more influential in whether people commit crimes or not. But the theory that religion deters crime has been a subject of serious academic discussion since at least the 1960s when criminologist Travis Hirschi and economist Rodney Stark proposed the "hellfire hypothesis." The hellfire hypothesis is that individual criminal behavior is deterred by the threat of supernatural punishment, and law-abiding behavior is promoted by the promise of supernatural rewards.[29]

A major review of research on this question began by noting that "existing evidence surrounding the effect of religion on crime is varied, contested, and inconclusive, and currently no persuasive answer exists as to the empirical relationship between religion and crime." Precisely because of the mix of results and interpretations, that study's goal was to re-examine all sixty studies of religion and crime that had been carried out across the United States between the 1960s and 2000, and to conduct a statistical meta-analysis that pools all of the findings together and tests the strength and direction of the relationship. The researchers found that across this wide range of studies, representing a variety of approaches and methodologies, the overall effect is that religion *does* seem to matter: "Religious beliefs and behaviors exert a moderate deterrent effect on individuals' criminal behavior."[30]

Religion is not a reliable deterrent of crime. Crimes are committed and prisons are filled by believers and nonbelievers alike, and crime is a problem in both secular and religious societies. Nevertheless, the evidence suggests that people with stronger religious beliefs are somewhat less likely to commit crimes. Clearly, it is not *only* the law that affects people's behavior—both for good citizens who are deterred and the criminals who are not. There is something else going on as well. Whatever earthly punishments may be understood to lie in store for us, people tempted to do wrong often are (or become) concerned that they will face supernatural punishment as well. Supernatural punishment is not the only stick available for enforcing cooperation, but two sticks are more powerful than one.

Secular and supernatural punishment may each be able to establish and maintain cooperation on their own. But in combination they reinforce each other, and indeed the total effect may be stronger than the sum of their parts. There are interactions by which secular and supernatural mechanisms of punishment may empower each other. Even where only secular laws apply, for example, they often have legalistic foundations rooted in the local religion—which adds to their salience for the people that must adhere to them. In the United States, certain events are still described and conceptualized as acts of God, people continue to swear on the Bible in court, and many moral principles are derived from Christian doctrine. Societies may be able to get along O.K. without religion, but it lends an almighty hand.

IS SUPERNATURAL PUNISHMENT A TOOL OF THE ELITE?

In the previous chapter I made the argument that individuals profit from beliefs in supernatural punishment because it helps them to avoid the costs of social transgressions. And now I have made the argument that societies at large also profit from beliefs in supernatural punishment, because it helps them to reap the benefits of cooperation. Clearly, these two effects reinforce each other. Moreover, the same belief helps both individuals *and* the group, reducing the usual tension between the interests of each. In this case, individual advantages beget group advantages and vice versa. As a result, supernatural punishment beliefs may be swept along quite rapidly by selection, as it is acting in the same direction at both the individual and group levels.

But this raises an important question: Does everyone benefit to the same extent? Karl Marx would likely mount a significant objection. Religion can, of course, be seen to improve the unity and productivity of a group of human beings, but to whom does the benefits of this unity and productivity fall? Or who is it helping most? In short, who stands to benefit?

People higher up in the social hierarchy may benefit much more than those at the bottom—from religious belief and practice in general, and from supernatural monitoring and punishment in particular. First, the advantages of increased cooperation may flow disproportionately to the top. Beliefs in supernatural punishment may sometimes, therefore, work as a tool of the elite, who impose or encourage beliefs in a punishing god (rather than other possible types of belief), the better to extract compliance from laypeople for their own benefit.

Second, if the masses believe God will punish violations of pro-scribed behavior, then the job of leadership itself becomes easier. Rulers don't have to watch over everyone's shoulder or dish out Hammurabi-like punishments, and may also enjoy a religious justi-fication for their authority, firming up their position as well as filling their coffers. In some societies there are also direct Darwinian ben-efits to religious leadership. When a shaman detects problems in the supernatural realm, the solution is sometimes held to be sex with the shaman. Indeed, male control of female behavior has been argued to be a widespread aspect of religion. While this may often appear to benefit men as a whole, in polygynous societies the benefits will accrue disproportionately to those at the top.[31]

Note that wherever elites have found it expedient to invoke super-natural punishment to promote obedience or conformity among lay-people, it offers further evidence that supernatural punishment is an effective method of achieving cooperation. This is, firstly, because if it were counterproductive, they would presumably not do it. Secondly, because it suggests that *even people with the power to administer sanctions themselves* find it necessary or advantageous to have gods do it as well, or instead. The authority of human leaders may not merely be justified by religious beliefs, but actually made more effective with the help of supernatural agents.

On the other hand, however, perhaps people lower down the hierarchy benefit *more* from beliefs in supernatural punishment, if they are the ones most vulnerable to the costs of social transgres-sions. The downtrodden caught flouting laws are likely to be pun-ished under the law. Elites caught flouting laws are more likely to get away with it (especially if the authorities are corrupt). The caution-ary mind-guard of belief in supernatural punishment might thus be more useful for a pauper than a prince. One could also argue that the propensity to cooperate is especially important to the everyday lives of ordinary people, because they must rely on trust and trade to get things done, whereas elites can use bribery, power, and coercion. Finally, a relative advantage to the lowly might arise indirectly if there are significant costs to leadership (in addition to its potential benefits). These costs may be in part borne by leaders to the advantage of followers.[32]

The opposing effects outlined above may cancel each other out, leaving both leaders and followers to benefit from religiously inspired coordination, cooperation, and increased productivity. For example, although Inca society was rigidly hierarchical and the emperor held his land "in an iron grip," the system appeared to work for both elites and laypeople. According to Hultkrantz, "everything, in practice as well as in theory, was done for the public good." In Christianity, and other major world religions too, conceptions of supernatural punishment often focused on being "fair," and did not rain down without constraint from on high: "The *lex talionis* of Ex. 21: 23–25 [an eye for an eye] is not the expression of a vindictive spirit. Rather it assures an even justice (the rich and the poor are treated alike), and a penalty proportionate to the crime." But what do the broader data suggest?[33]

At first glance, religion appears to extend the power and reach of leaders—in both this world and the next. Leaders in numerous historical societies around the world have been elevated to the status of gods, or were said to become gods after their death, as with Roman Emperors such as Julius Caesar. In some Polynesian cultures the souls of chiefs go to a special heaven, where they acquire godly status. They may even acquire godly status on Earth, enabling them to more effectively extract compliance. During the annual *Makahiki* ceremony in ancient Hawai'i, the high chief "trades places" with the god of agriculture, *Lono*, as part of a seasonal fertility rite. A special *lei* normally worn only by chiefs is placed around the neck of the god, and then the chief supposedly stays in the temple while the god (in the bodily form of the chief) "proceeds around the island for a month receiving tribute and taxes."[34]

But what if we turn our attention to the small-scale foraging societies that provide a window onto our ancestral past? Perhaps the single most striking thing about these societies, especially as opposed to the chiefdoms and civilizations that came after them (such as the Hawaiians and Romans), is how egalitarian they were. Although there are differences in power and reproductive success among individuals, one of the defining features of these societies is a lack of a leader or any strong hierarchy at all. There are individuals who are recognized as especially skilled in certain tasks—hunting, making tools, or indeed in religious matters, sometimes presided over by shamans. However, on the whole it was a level playing field. Here, therefore, in the socioecological context in which religious beliefs and behaviors arose, supernatural reward and punishment would have applied equally to everyone. There were no rulers who could stand above it, or benefit from imposing it upon their followers. For this reason, it seems likely that religious beliefs and behaviors—at least in our deep ancestral past—must have had positive adaptive advantages for the masses. Religion was not a tool of some (nonexistent) elite. This lends additional support to the idea that religion has its origins in adaptive advantages to individuals.[35]

Supernatural punishment, then, is likely to have emerged and evolved independently of any influence of leadership and hierarchy. Even as religion developed in more complex societies, there are reasons to believe that the system only worked if commitment to religious beliefs was genuine among leaders as well as followers. As Joseph Bulbulia notes:

> Religious agents remain exposed to information arrayed to deceive, defraud, and manipulate them. However, because religious organizations are generally functional, and because religious authorities will generally have survived the costly testing of their commitments, the risks for manipulation are significantly reduced. Clearly, the risks are not eliminated. The guru intent on a harem will always remain a threat, but ... given the entrenchment of religiosity in our psychological design, we think that the best explanation for psychological dispositions to structured religious pre-commitments is that they best served the biological interests of those who were prone to them over the vast epochs of human evolution. The evidence suggests they helped to make large scale cooperative exchange an evolutionary reality deep in the human lineage.[36]

So perhaps we should not be too cynical about power and leadership, even in the larger societies that came later. If large-scale social order is to be achieved, then someone has to lead, and others have to follow. And for this to succeed, any given follower does not necessarily have to suffer a net disadvantage. Indeed, for society to succeed at all followers may have enjoyed a net benefit of being in a group, even if (a few) leaders gained more. The advantages of society are many, and may accrue to any or all of its members. Even where it is clear that the average individual incurs some cost from compliance with religious norms, the resulting benefits may be greater. At least, the *perception* that this is the case is common, and may be built into religious beliefs themselves even if it means bolstering rather than breaking down social structures. Sociologist Max Weber saw this demonstrated most remarkably in the caste system in India, where

> any effort to emerge from one's caste, and especially to intrude into the sphere of activities appropriate to other and higher castes, was expected to result in evil magic and entailed the likelihood of unfavourable incarnation hereafter. This explains why, according to numerous observations on affairs in India, it is precisely the lowest classes, who would naturally be most desirous of improving their status in subsequent incarnations, that cling most steadfastly to their caste obligations, never thinking of toppling the caste system through social revolutions or reform. Among the Hindus, the Biblical emphasis echoed in Luther's injunction, "Remain steadfast in your vocation," was elevated into a cardinal religious obligation and was fortified by powerful religious sanctions.[37]

Religion may not, therefore, only serve to keep people in their place (for good or ill), but it may also keep them wanting to stay there. Whether they are the recipients of duplicity or dividends is not entirely obvious, because while social climbing may be good for the individual, it may be bad for the group.

A final aspect of religious leadership is neither its role in coercion nor conspiracy, but in providing an *example* to follow. History has many instances of charismatic or otherwise remarkable individuals who, by their own prosocial or self-denying acts, encourage others to do the same or at least to make efforts toward their ideal—Moses, Jean d'Arc, Ghandi, and Martin Luther King are notable examples. Without such leaders, whole movements or societies might never have succeeded or survived. This more positive social role of leadership has gained prominence in evolutionary theory because of new work identifying imitation as highly adaptive in the context of cultural evolution.

Much of the recent research in evolutionary psychology on how and what people learn has focused on the role of prestige and the utility of copying exemplary individuals. Anthropologist Joe Henrich suggests that religions are special in this regard because when religious leaders (such as ascetics or martyrs) make costly sacrifices, these acts in themselves add credibility to their avowed faith—why would they do it if they didn't genuinely believe? Others are, as a consequence, more likely to follow. With religion, actions speak louder than words. These "credibility enhancing displays" may be impressive when performed by anyone, but are especially effective if performed by someone in a position of leadership, since people can see that they practice what they preach. This can give rise to a process of positive feedback, since such acts spread the message to additional believers who may follow suit, generating a "self-reinforcing loop" that spreads and stabilizes the system of beliefs and actions over time. Henrich suggests that via this process, even apparent setbacks for a religion may counterintuitively have aided their cause. For example, the persecution of Christians by the Romans may have inadvertently helped the religion to spread because followers were martyred for their beliefs, providing an impetus for others to adopt, increase, and pass on the faith. The crucifixion of Jesus Christ may represent the zenith of this effect.[38]

To summarize, supernatural punishment is not a ruse of the elite. It is a deterrent that works precisely because it places authority and ultimate power way beyond what any mere mortal is capable of—whether peasant, priest, or potentate.

WITH WHOM TO COOPERATE? GROUP BOUNDARIES AND BIASES

An important clue to the tight connection between religion and society is how religious groups see themselves in relation to other groups. Religion is often seen as binding together a collection of people in specific opposition to a sea of unbelievers—whether of different religions, different sects within the same religion, or atheists. People define their group memberships according to multiple different categories (family, firm, church, nation, and so on), but religion often appears to have a special prominence. As David Sloan Wilson says, religion is "the groupiest thing around." A critical concern for religious groups has often been to distinguish themselves from others. But not usually as equals.[39]

A recurring theme in religious narratives is the idea of being "chosen people." A given society generally finds that of all the people on earth, God or gods are particularly interested in *them*. This is common across indigenous, historical, and modern societies. It is clear, for example, in the Hebrew Bible, which specifically focuses on the Israelites as a chosen people. But the concept of a chosen people is not just a vestige of the Old Testament. For example, Jesus himself says: "It was not you who chose me, but I who chose you" (John 15: 12–16). This is sometimes interpreted as a metaphor, or a historical perspective, rather than something fundamental about Christianity itself. But if we trace the development of religious groups and the competition between them, it becomes obvious that the conception of gods and God changes as it tracks changes in the membership of the relevant group, and most notably, as groups expand. In one well-known example, Islam partly spread via trade networks, with common religious beliefs and practices helping merchants to build the trust necessary to do business over long distances and time lags between desert voyages. More recently, Robert Wright argued that Paul's expansion of Christianity operated in part through exploiting the network of Christians in towns across the Middle East, in which people of shared faith could seek each other out for help.[40]

The tight bonds of religious groups are also powerfully exemplified in attitudes toward those who *leave* the religion. In Islam, rejecting

God is regarded as the worst possible thing one can do. For this reason, many fundamentalist groups are focused on the apostasy of west-
ernized, secular governments within the Muslim world, and see these
as far more contemptible than distant Christian nations who, even if
they are disliked, are at least not apostates.

The idea that supernatural agents are specifically focused on the
in-group is exemplified across the spectrum of human societies. In
Hartberg and colleagues' study of how religious sanctions help the
management of common pool resources among small-scale indigenous
groups, the authors found that "unlike high gods, nature and ancestral
spirits are generally provincial phenomena with relatively narrow spheres
of influence and restricted domains of moral interests." In the middle of
the spectrum, larger communities tend to be brought together by moralizing gods. And at the far end of the spectrum, the spread of the major
world religions has only been possible by extending the circle of territory
and people who are permitted or obliged to join the group, often culmi-
nating in a single, all-powerful God. Wright found this pattern to repeat
itself multiple times in the development of the Abrahamic religions. The
broader the set of peoples who joined together under the same political or economic unit, the more powerful and all-encompassing gods
became. After all, religious proscriptions and punishments have little
relevance to those outside the group. This distinction is also a feature of
both modern and indigenous religions. The Hartberg study found that
"when faced with the uncertainty of whether a distant group can be
trusted to cooperate, individuals may not find it plausible that breaches
of that trust will really be punished by the other group's ancestral spirits." In short, cooperation breaks down beyond the boundaries of the
group—or, more specifically, beyond the boundaries of shared belief.
All of these features of religious groups are interesting in themselves.
But they are especially interesting in the light of the significant obstacles
to cooperation we have explored. Cooperation relies on a clearly demar-
cated group of people—otherwise free-riders cannot be held to account,
and individual sacrifices cannot reliably return benefits to either the
individual or their group.[41]

SUPERSIZE HIM: FROM TINY
TO MASSIVE SOCIETIES

In the societies of our evolutionary past groups were small, leaders were absent, and supernatural consequences came from local spirits and ancestors rather than gods. How did all of these things get reversed? How did we wind up with massive societies, rigid social hierarchies, and supernatural punishments delivered by powerful gods and eventually, in some societies, a single monotheistic God?

I suggest that amidst a lot of complex influences, the fundamental dynamic at work was a simple one: cooperation. The problems of achieving cooperation, even in small groups, were laid out in Chapter 2. However, the challenges to cooperation are one thing in small, relatively stable groups, where everyone knows each other, tend to be related, and reciprocate over lifetimes. But achieving cooperation is quite another thing in massive, shifting populations of anonymous strangers, with divergent interests and incentives, facing multiple collective action problems and demanding large-scale public goods. Here was a challenge to cooperation par excellence. How did we overcome it?

There are many contributing factors, but not least of these may have been religion. Weber noted in *The Sociology of Religion* that "there is no concerted action, as there is no individual action, without its special god. Indeed, if an association is to be permanently guaranteed, it must have such a god. Whenever an organisation is not the personal power base of an individual ruler, but genuinely an association of men, it has need of a god of its own . . . It is a universal phenomenon that the formation of a political association entails subordination to its corresponding god." The data seem to corroborate Weber's basic insight. Anthropologist Guy Swanson's famous comparison of fifty indigenous societies around the world found that as societies increased in political complexity gods were not necessarily more common or important, but they became more moralizing. In other words, as societies became larger and more structured, the type of gods people envisaged became more involved with their moral conduct. Perhaps, in other words, big societies were not possible without big gods, as social psychologist Ara Norenzayan has argued.[42]

But before we consider how the transition may have ended, let us consider how it began, and the specific role that religion may have played. If we trace the evolution of our species and the emergence of religion, we find some important convergences. For example, in the previous chapter I argued that theory of mind and complex language marked a critical milestone in our evolutionary journey, and it was only at this point that supernatural beliefs became important (indeed, even possible). What can anthropologists tell us about the dates of these events? Looking at a variety of fossil and associated evidence, some suggest that speech originated around 50,000 years ago. That would tie in to some of the earliest cave paintings and widespread burials with grave goods, thought to be intended for an afterlife. The origins of theory of mind are harder to nail down, but some researchers have suggested that the emergence of theory of mind roughly corresponds with the emergence of grave goods as well. The other clue is the appearance of symbolic—rather than purely functional—objects such as jewelry. The earliest known appearance of jewelry is 82,000 years ago, found in limestone caves at Grotte des Pigeons, near Taforalt in Morocco (some 40,000 years before any such artifacts turned up in Europe). They were made from shells, transported some distance from the sea, and painted in red ochre. Surface patterns suggest they were worn as necklaces or bracelets. Both grave goods and symbolic objects suggest a range for the emergence of theory of mind of somewhere around 50,000 to 100,000 years ago. This is significant because with theory of mind and language developing around the same time, we have the building blocks not only for the perception of unseen agents and supernatural causes and effects, but also the tools to share these experiences and generate cultural narratives. It also suggests there has been sufficient time *since* then for natural selection to hone our propensity for supernatural beliefs. If such beliefs helped us, they may have become stronger, or adapted into different forms. Selection may even have shaped the kinds of supernatural agents and belief systems that tended to work well in suppressing selfish behavior among individuals and promoting cooperation in groups.[43]

Much later, there was another significant landmark in the road to modern life—the development of agriculture around 10,000 years ago. This led to the formation of large, sedentary, and urbanized societies, facilitated by the ability to grow and store surplus food, live in dense and permanent towns, and support divisions of labor. This transition from small to large societies unleashed a sea change in human history, with significant consequences for social organization and behavior. It has often been assumed that complex religions came *after* societies became complex, but there is some interesting evidence suggesting that it happened the other way around: Changes in religion itself may have been an engine that propelled society forward. It opened up new opportunities and increased a society's capacity and productivity. One way this may have happened is by increasing the levels of coordination and cooperation that were possible. And as this book has argued, any such strides forward would have received a terrific boost from a belief that supernatural agents monitored and punished people who violated social norms.[44]

But the discontinuity from foraging to agriculture in our evolutionary history suggests two different ways in which supernatural punishment may have worked. An emerging debate on the role of supernatural punishment in human society has divided into two versions. First, there is punishment avoidance, as outlined in Chapter 5, in which the threat of supernatural punishment *suppresses individuals' selfish behavior*. Second, there is cooperation enhancement, as outlined in this chapter, in which the threat of supernatural punishment *increases groups' cooperative behavior*. Clearly, these two mechanisms reinforce each other and may thus be two sides of the same coin. However, they have interesting differences in their causes and consequences. Punishment avoidance invokes supernatural punishment as a solution to the rising *costs* of selfishness (supernatural punishment comes into play because the risk of real-world costs from social transgressions is *high*). By contrast, cooperation enhancement invokes supernatural punishment as a solution to the rising *benefits* of selfishness (supernatural punishment

comes into play because the risk of real-world costs from social transgressions is *low*). Punishment avoidance and cooperation enhancement
may appear, therefore, to offer two contradictory proposals. However,
the reason they have these opposite effects is because they apply in different social contexts—contexts that were dominant at different times
in human evolutionary history. When humans lived in *small* groups,
transgressions were more likely to be detected (increasing the costs
of selfishness to the individual, and thus making transgressions less
attractive), but when humans began to live in *large* groups, transgressions were much less likely to be detected (making them more attractive, and thus increasing the costs of selfishness to the group).[45]

Punishment avoidance and cooperation enhancement can thus
be seen to be complementary outcomes of the same set of beliefs,
but each coming to prominence in a different socio-ecological context (see Table 7.1). Punishment avoidance is primarily about how
a fear of supernatural punishment emerged in human evolution,
when reputations became threatened by theory of mind and complex language, sometime in the Pleistocene (1.8 million to 10,000
years ago). Cooperation enhancement is primarily about how a fear
of supernatural punishment developed in more recent, large, anonymous societies, when the reputations of strangers could not be so
easily tracked, sometime in the Holocene (10,000 years ago to the
present). With this perspective, punishment avoidance and cooperation enhancement are not at any time mutually exclusive theories,
but insights about how belief in supernatural punishment can have
positive effects in different kinds of social settings. Both appear to
function in historical and recent societies, and they reinforce each
other.[46]

The problem with cooperation enhancement on its own is that it
suffers from the free-rider problem, and thus requires group selection
for cooperative traits to survive and spread (because individuals that
cooperate benefit the group at the expense of their own self-interest).
As we saw in Chapter 2, group selection faces significant hurdles if it
is to work. However, it may have become an important dynamic in
more recent human evolution, because certain situations have emerged

Table 7.1 Key components of supernatural punishment in two major eras of human evolution: the Pleistocene (1.8 million to 10,000 years ago) and the Holocene (10,000 years ago to the present). The increasing size of societies was accompanied by a significant shift in the type and function of supernatural agents.

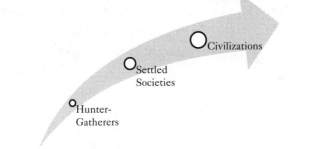

	Pleistocene	Holocene
Population size	Small	Large
Key problem	Maintaining reputation in small groups	Maintaining cooperation in large groups, and intergroup competition
Costs of being selfish	High (free-riders are obvious)	Low (free-riders melt into the crowd)
Role of supernatural punishment	Punishment avoidance	Cooperation enhancement
Primary mode of selection	Individual selection	Group selection
Precipitating events	Evolution of theory of mind and language	Agriculture, food surpluses, settlements
Primary types of supernatural agents	Spirits, ancestors	Moralizing gods, God

in which the hurdles for group selection are lowered. First, if human societies experience severe intergroup *competition or conflict*, then whole groups can directly out-compete, damage, and even replace each other, and this process of selection may become stronger than selection operating at the individual level. If actual intergroup *warfare* was important in the Holocene, then cooperation enhancement would fit with biologist Richard Alexander's hypothesis that moralizing gods became vital to maintaining social cohesion and collective action when groups of burgeoning size began to fight—and needed to deter and survive—large-scale wars. Second, if human societies develop *cultural* ideas and practices that vary among groups, and have differential effects on the groups' success, then the survival and spread of cultural traits rather than genes—religious beliefs and behavior included—may come to depend on selection acting at the level of groups.[47]

CULTURAL EVOLUTION: SPEEDING UP THE SUPERNATURAL

This book has focused on biological selection. If a given physiological or psychological trait helped our ancestors in human evolutionary history, then all else equal, the beneficial trait would have been favored by natural selection, and the genes responsible would have increased in frequency and spread in subsequent generations. However, this is not the only way evolution works. And these other ways may mean supernatural punishment has had an even greater influence on human society.

Darwinian selection takes place whenever three features are in place: (1) variation in some trait; (2) greater success of some variants of the trait than others; and (3) replication of successful traits. While we tend to think of this process operating on genetic, organic matter, those three features can in fact occur in all sorts of domains. Genes can vary, have different outcomes, and spread. But so can other things. A range of entities that can be subject to the process of Darwinian selection include (at least in principle) genes, individuals, groups, words, fashions, machines, robots, companies, organizations, strategies, and ideas. The underlying process—variation, selection, and replication—does not change. Only the *unit* and the *agent* of selection change: the unit is the thing being selected, and the agent is the filtering process that preserves some variants over others—for whatever reason. In effect, information is passed on in a different form, and by a different editor, but *how* it does so is broadly the same. In biology, information is coded in genes. But, of critical importance to our discussion of religion, information can also be coded in our brains. Therefore, an idea—a piece of information—can spread from one mind to another, just as genes can spread from one body to another. With multiple ideas whizzing around, some of them are likely to be ineffective or detrimental, and will be abandoned or forgotten. But others will be good ideas, and will stick around and be imitated or copied by others, and thereby spread. Thus we have a process of Darwinian selection, but of ideas, not genes. It is called *cultural selection*.[48]

Evolution can therefore include the *cultural transmission* of ideas and practices, as well as the *biological transmission* of genes. If we are to understand such a complex, multifaceted social phenomenon as religion, then this cultural evolutionary perspective is vital. As Richard Dawkins notes, "whatever theory of religious evolution we adopt, it has to be capable of explaining the astonishing speed with which the process of religious evolution, given the right conditions, can take off." Cultural selection is interesting precisely because it operates at breakneck speed. With biological selection, we have to wait for multiple generations to see which variants survive, spread, or die out. With cultural selection, we can witness ideas rising and falling by the day.[49]

But cultural selection is important for another reason: It operates powerfully on *groups*. It is at the level of groups that ideas tend to dominate, diffusing quickly within groups but more slowly between them—indeed, group-specific traits often serve as markers to demarcate the group itself, with individuals adopting and displaying the trait to signal their group membership. These are precisely the conditions required for cultural group selection to work. The trait must vary among different groups, but each group's constituent individuals must tend to share the same trait. This generates between-group disparities in cultural traits, and between-group disparities in any effects these traits may have on a group's success. Hence, cultural traits tend to vary among groups rather than among individuals, and this sets the scene for the selection of ideas at the level of groups.

But how does that work in practice? Cultural selection can occur by copying, competition, or conquest. Imagine some problem, such as how to build a boat. Group A knows how to do it, but group B does not. How could knowledge of how to build a boat spread? First, group B can copy how group A does it. If so, one idea for solving the problem has spread from one group to another but each group's composition and size stay the same (here, the idea spreads by observation or communication). Second, group A may do better than group B (perhaps because the boats themselves somehow allow them to be more efficient or productive). If so, group A spreads while group B dwindles (here, the ideas spread with the spread of the people themselves). Third, group A can conquer group B, bringing with them or imposing their boats and boat-building expertise (here, the ideas spread from the victors to the vanquished of the conquered territory). Over time—and in all three scenarios—the real determinant of whether the new idea is successful or not will come from the rigors of the environment. If it doesn't work in a given social or physical setting, it won't last and will bow out to alternatives. If it does work, it will survive and spread.

Some people see this process of cultural evolution as so significant that it represents a new step in the so-called major transitions of life. The major transitions of life are significant breakpoints in evolutionary history in which individual elements combined to form larger collectives—from molecules to cells, from cells to bodies, from individuals to groups. It is *other* groups that represent the next frontier. The role of cultural group selection acting on humans, plus our ability to adapt the environment to ourselves rather than us having to adapt to the environment, have become vital players in our evolution.[50]

So while it was largely genetic selection operating in the small-scale human groups of our evolutionary past, cultural selection became influential when human groups became larger, technologically sophisticated, and conquering. However—and this is an important point—the things we tend to create, like, and copy are themselves heavily shaped by our underlying evolved genetic traits and propensities. Cultural selection, therefore, is not divorced from our biology. It operates within boundaries set by our genes, and indeed relies on those genes for ideas and practices to develop in the first place. Our underlying biology therefore gives us predictions for what kinds of cultural traits are likely to emerge, succeed, or fail. While cultural selection may therefore be responsible for many variations in the precise content of religious beliefs and behavior, the genetic and cognitive architecture underlying those beliefs and behaviors mean that they will tend to follow certain patterns rather than others. It is precisely these recurrent, cross-cultural, and pan-historical patterns that I have endeavored to show in the case of supernatural punishment—remarkably diverse in form, but common to religions across the globe and throughout history. Though it may have long been favored by natural selection, cultural selection clearly powered forward the role and reach of supernatural punishment among human minds and societies. It was an idea that worked, spread, intensified, and prevailed. And nowhere was it more important than in the development of human groups into societies, civilizations, and nations.[51]

CONCLUSIONS

In the previous chapter, we explored the adaptive advantages of belief in supernatural punishment for *individuals*. This was counterintuitive. At first glance religious beliefs and behaviors seem to impose costs that act against self-interest and Darwinian fitness. Yet we discovered that in fact a belief in supernatural punishment can help to *reduce the costs of selfishness* that an individual pays in retaliation and reputational damage, and opens the way to greater gains from mutual cooperation with other individuals. This chapter highlighted how the benefits of supernatural punishment for *groups* and *societies* as a whole are greater still, because if self-interest is suppressed and cooperation is catalyzed, then we can all reap benefits far greater than the sum of our individual activities. We can not only trade favors or lend a hand to a neighbor, we can build irrigation systems, roads, and cities.[52]

In early human societies, several basic cooperative activities were crucial to survival—hunting, food sharing, divisions of labor, construction, collective defense, warfare, and so on. Even here, however—or perhaps *especially* here, in the case of costly and dangerous activities such as hunting and warfare—cooperation is always open to exploitation by those who cheat the system. While social cooperation may have large, obvious, and even urgent advantages, there is always an Achilles heel: free-riders, who reap the benefits of collective action without cooperating themselves. As soon as humans found themselves in the rapidly changing environment of large, anonymous societies, this Achilles heel became magnified. People were harder to monitor. Reputations became harder to sully. Cheats became harder to catch. And punishments became harder to administer. Worse, humans had to deal with this super-free-rider problem with the cognitive tools designed for life in a much simpler environment of small, familiar kin groups. To keep the creaking ship of society afloat, supernatural punishment would have become an ever more important tool. Rather than human nature adapting to the environment, societies needed to adapt the environment to human nature. One important feature of the environment available for use was a pantheon of supernatural agents. In a process of rapid cultural evolution in the recent human past, gods became more common, more moralistic, more powerful, and more fearsome.

The authors of the Federalist Papers noted that "if men were angels, no government would be necessary." Unfortunately, men (and women) are not angels. There is always the temptation to engage in selfish behavior, to exploit other people, or to enjoy the benevolence of others without contributing anything back. Not everyone does so, but you only need a few free-riders, or a lot of people free-riding a little bit, to undermine cooperation. Numerous experiments, mathematical models, computer simulations, and field studies demonstrate that free-riding can easily lead to a breakdown of cooperation—even if most people are initially prepared to cooperate, and even if they are prepared to forgive defectors at first. The ever-present problem of free-riders remains a fundamental problem for game theorists—and indeed for governments and policy makers throughout history and today—and it has led to years of research to uncover the conditions under which cooperation can evolve and endure in spite of them. Especially in the era before governments existed, any trait that helped to solve this problem would have conferred significant advantages to the individuals and groups that discovered it. Human nature makes us especially sensitive to the idea of being watched and punished for our actions. I have argued that, especially when cemented and amplified by cultural and religious narratives, exactly such a belief—a belief in supernatural punishment—offered a remarkably effective solution to the problem of cooperation, offering a level of surveillance, dependability, and punishment that no human agent or institution could match.[53]

People continue to believe that what goes around comes around. If these beliefs suppress our self-interest and promote our cooperation, then it may be good for us as individuals, but it may be even better for society at large. In fact, it may have been a prerequisite for societies to *become* large in the first place, or even to become societies at all. Supernatural punishment may seem like a destructive, negative force, wielding sticks where carrots could be offered instead. But a formidable deterrent may have been exactly what was needed to solve the pervasive problem of free-riders that stood in the way of society. God was a game changer. But now that we have got here, do we need God anymore?

CHAPTER 8

THE WORLD WITHOUT GOD

People say we need religion when what they really mean is we need police.
—*H. L. Mencken*

In December 2004, a bizarre item appeared for sale on eBay. A woman named Mary Anderson was looking for someone to buy her father's ghost. She was ready to part with this item because her six-year-old son, Collin, was so afraid of the ghost that he would not walk around the house alone. Collin told his mother: "He was mean. His ghost is still around here!" How would eBay help? It was a ruse to assuage her son. Once the ghost had a new owner, she explained to him, it would go to a new home. Mary Anderson asked whoever the buyer might be to write a letter to Collin confirming receipt and "letting him know that he's there with you and you're getting along great." Nice idea. But of course no one would put up hard-earned cash to buy a ghost, right? Wrong. There were 132 bids, with the winning offer a staggering $65,000.[1]

In the Internet age, we are faced with a kaleidoscope of supernatural ideas, superstitions, cults, new age beliefs, and religious groups, along with more or less direct access to billions of people. However weird or unusual our beliefs, in such a large population it can be easy to find others who share them. For example, one can join the First Electronic Church of America (FECHA), the Church of the Rainforest, or the First Online Church of Bob. For different kinds of otherworldly pursuits, we can plug into communities of tens or hundreds of thousands in massively multiplayer online role playing games (MMORPGs), such as the million active members of Linden Lab's *Second Life* online community. There are also the growing online activities of existing mainline churches, and they are merging with social networking sites to become an active part of people's everyday lives. Combine that with the real-world urban melting pot of people and ideas, clubs and cults, religions and churches, and we have an embarrassment of riches for spiritual fulfillment. Anything, it seems, is possible.[2]

One might think that if it wasn't already, religion has become far too varied and complex to allow us to speak of it in terms of universal human nature or biological underpinnings. But I suggest that the opposite is true. The diversity of beliefs may be staggering, but it is not infinite and not random. Rather, there are certain *kinds* of things that people tend to believe rather than others. And there are certain *patterns* in the way we go about putting beliefs into practice. Human brains have architectural structures and neurological processes that, while allowing considerable freedom, predispose us to pay more attention to some things and less to others—and not least when it comes to religion, as cognitive scientists have demonstrated.

One of the things we pay a lot of attention to is the supernatural consequences of our actions. And we have seen how this has been a concern for human beings across the world, throughout history, and apparently deep into our evolutionary past. But the world is changing. Rapidly. What are the implications of our supernatural punishment mindset for how we, as individuals, go about our lives today? What are the implications for human society at large? And what about the future? After many tens of thousands of years surrounded by supernatural agents with an inclination to intervene in our lives, how will human beings and human societies cope with losing this part of our evolutionary heritage as the world—or at least the West—becomes increasingly secular?

THE ATHEIST THREAT: DO WE NEED GOD TO BE GOOD?

People often argue that religion is essential for moral behavior. Many religious doctrines, leaders, and adherents claim that without religion, individual moral standards and social harmony will break down.

Influential opinion leaders even use the threat posed by atheists as a tool to motivate their followers. This has been a strategy in western civilization since the origins of the concept, as religious historian Jan Bremmer reminds us: "Greeks and Romans, pagans and Christians, soon discovered the utility of the term 'atheist' as a means to label opponents."[3] The slur continues today. For example, in the United States, Pat Robertson has claimed that the result of society without religion will be "tyranny." Evangelical and influential political commentator Jerry Falwell declared that 9/11 was caused by secular Americans angering God. Conservative columnist and commentator Ann Coulter suggested that societies without a sufficient understanding of God will slide into slavery, genocide, and bestiality. And Fox News anchor Bill O'Reilly has warned that a society without God will lead to anarchy and crime.[4]

Humanists and New Atheists reply, of course, that this is absolute rubbish. The key counterargument is that obviously not all atheists are criminals or free-riders, and not all secular nations are in anarchy, cruel to their citizens, or dangerous to other states. Similarly, not all religious people are law-abiding, and not all religious states are benign. So even if religion is generally a force for good, it is evidently not a necessary or sufficient condition. Others have tested the proposition scientifically. An experimental study by psychologist Marc Hauser and philosopher Peter Singer found that there was no difference between atheists and believers in how they responded to common ethical dilemmas. Clearly, there is no simple connection between religion and moral behavior among either individuals or societies. Atheists and humanists have been especially keen to advocate this view—and even advertise it, with publicity campaigns to spread the message. When Pope Benedict XVI visited Britain in 2010, the Humanist Society of Scotland posted giant billboards declaring "TWO MILLION SCOTS ARE GOOD WITHOUT GOD". The American Humanist Association had an ad campaign on Washington, DC buses that asked: "Why believe in a god? Just be good for goodness' sake."[5]

The bigger issue, however, is that irrespective of logic or data, a lot of people continue to *believe* that religion is necessary for morality, whether it is or not. According to a 2008 poll by the Pew Forum on Religion and Public Life, which asked people "Is it necessary to believe in God in order to be moral?," 57 percent of Americans said "yes" (and 41 percent said "no"). The "yes" figures were 30 percent in Canada, 22 percent in Britain, 17 percent in France and, the lowest figure of all, only 10 percent said yes in Sweden. Outside the West, the figures were even higher than in the United States: 66 percent in India, 82 percent in Nigeria, 84 percent in Turkey, and 99 percent in Egypt. Whatever the psychologists and philosophers say, in large parts of the globe a majority of people think religion is crucial for morality. This belief is likely to affect their attitudes, votes, and behavior—whether it should do or not.[6]

People's concern is not necessarily that without God all hell would immediately break loose, but rather that moral standards would slip, and people would be more self-interested and less cooperative, generous, or helpful to others. New research suggests that the perceived problem with atheists is that whatever their true nature, people don't trust them. As Will Gervais, Azim Shariff, and Ara Norenzayan put it in the opening of their study into attitudes toward nonbelievers, "atheists are among the least liked people in areas with religious majorities (i.e., in most of the world)." For example, atheists are the "least desirable group" for people's son or daughter to marry. And only 45 percent of Americans would vote for an atheist presidential candidate, even if they were qualified in all other respects. One study found that the level of contempt for atheists places them on a par with rapists. Other studies found that—get this—even atheists do not trust other atheists (it takes one to know one, perhaps!). The message is clear that at least in North America, where most of these studies have been done, atheists are not seen as reliable contributors to society.[7]

The point of all this is to underline something critical to the argument of the book. Atheists are not just seen as a problem in and of themselves, they are explicitly seen *as a problem for achieving and sustaining social cooperation*. In the game theoretical terms of Chapter 2, they are assumed to be free-riders (or to be more likely to free-ride), even before they have had any chance to show whether they might be a free-rider or not. And it is not just a random prejudice against outsiders from another group. People do not *dislike* atheists, as such. Rather, they simply *distrust* them. The study by Gervais and his colleagues managed to obtain experimental evidence on exactly why this is the case. It seems to be precisely because people believe that without God watching over them, atheists will be less restrained in their selfish behavior. As they explained, "the relationship between belief in God and atheist distrust was fully mediated [that is, accounted for statistically] by the belief that people behave better if they feel that God is watching them." The reason people don't trust atheists is not because atheists are intrinsically different kinds of people, belong to a different group, or have different rules. Rather, it is because they are blind to the eye in the sky.[8]

We are faced with the empirical reality that many of us believe people will not act in the best interests of the community or society if they are released from an expectation of supernatural consequences of their actions. Especially interesting is that evidently, secular consequences—laws, police, fines, jails—are not perceived to be good enough on their own. This is telling given the book's thesis that supernatural punishment has a special leverage that secular institutions cannot match. People certainly seem to think so even if it is false. Neither religion nor government are infallible methods of deterring selfishness and promoting cooperation, but—at least in the mind of man—a combination of the two may be much more effective than relying on either alone, as Pierre-Joseph Proudhon envisaged in his 1808 painting of "Justice *and* Divine Vengeance Pursuing Crime" (Figure 8.1).

LOSING OUR RELIGION

Only within the last few decades have we—at least in countries in the West—started giving up the idea of God. The Big Brother of organized religion is being lifted from our shoulders as our upbringing, education, peers, and politics become increasingly secular. Science is

Figure 8.1 "Justice and Divine Vengeance Pursuing Crime," Pierre-Joseph Proudhon (1808). © RMN-Grand Palais/Daniel Arnaudet.

chipping in as well, providing ever more and ever better explanations for things that once had no explanation at all. The extent of the decline of religion in western states is remarkable. In Europe in particular, religion has more or less been banished from politics, is mocked in the media, and is increasingly absent among the population at large, while churches are being converted into houses, sports facilities, shops, even bars. If God is dead, he must be turning over in his grave. The 2010 Eurobarometer survey asked people if they believe there is a God, and while belief remains high in Italy (74 percent) and Spain (59 percent), it has dropped to less than half in many other countries, for example, to 44 percent in Germany, 37 percent in the United Kingdom, and 27 percent in France (the lowest was the Czech Republic, with 16 percent). While in the United States no presidential candidate could be elected to office if he was an atheist, in Britain or France, no leader could be elected if he was overtly religious. In many aspects of life in Europe, religion is already gone.

But the interesting question is not so much the spread of nonbelief, but its consequences. What, in these societies, is changing as a result? How will the decline in religious belief, and the demise of the deterrent of supernatural punishment, affect selfishness and cooperation now and in the future? German philosopher Friedrich Nietzsche wrote that "there is no small probability that with the irresistible decline of faith in the Christian God there is now also a considerable decline in mankind's feeling of guilt." If we feel less guilt, we will be less likely to worry about the bad deeds of our past and less likely to be deterred from committing them in the future. Atheists and humanists may well be "good," but how good? Are they *as* good? Are they *as* productive or *as* cooperative as their believing peers? How does this scale up at the level of societies as a whole? And would this change if *everyone* became an atheist? Perhaps a lack of supernatural punishment is not a problem per se, but where it remains present it can increase cooperation and flourishing beyond what would otherwise be possible. The study by Harvard economists Robert Barro and Rachel McCleary found that across countries, societies that are more religious have greater economic performance—and interestingly this relationship was driven by religious *beliefs* in heaven and hell, not just religious *behavior* such as church attendance. And Shariff and Rhemtulla's study found that crime rates are lower in countries with stronger beliefs in hell (rather than heaven). Such studies are notoriously tricky to interpret because of possible confounding factors and historical contingency—even if many are statistically controlled for. At the level of national comparisons, therefore, the jury is perhaps still out. But what we can say is that there is at least some evidence that weaker religious beliefs, and weaker beliefs in supernatural punishment in particular, are associated with negative social outcomes (which dovetails with the findings of the laboratory experiments we explored earlier). This is precisely what the supernatural punishment theory would predict.[9]

REPLACEMENTS FOR GOD

If God is in decline, what, if anything, is replacing supernatural punishment as a means to temper self-interest and promote cooperation? And what seems to happen in societies that have already "lost" their religion? This might give us an idea of what will happen in a world without God.

BIG GOVERNMENT

Modern governments are fulfilling many of the institutional functions of religion with laws, police, courts, and punishments, as well as security, welfare, charity, and a range of democratic ideals to believe in. There are even analogues of supernatural monitoring that are cropping up where God once stood. For example, CCTV and surveillance cameras are a ubiquitous feature of many urban environments. In the United Kingdom, there are estimated to be up to 6 million cameras on the streets—more than in all of China, and one for every eleven people. This, along with other surveillance paraphernalia, has created a so-called big brother society where we are subject to round the clock monitoring by the unblinking stare of government institutions. Catalyzed by the threat of terrorism, governments have also become rather heavy-handed with monitoring phone calls, e-mails, bank transactions, and other activities. Numerous private companies such as Google and Amazon are in the game too. Knowing your deepest desires and tracking your behavior is big business as well as big government. Revelations about such surveillance routinely raise the ire of vocal segments of the populace and political elite, but the rise of big data tracking us all seems an ineluctable aspect of modern life that we are going to have to get used to.[10]

These developments may seem unnecessary or futile, but policemen are expensive and populations are expanding. Moreover, criminals and terrorists have been successfully brought to justice with the help of such surveillance. Even if such successes are rare, if they are important enough, it may be worth the cost. The debate is what that cost is. How do we evaluate the intrusion of privacy that is imposed on everyone under the gaze of millions of cameras and other devices running day and night around the world? Oddly enough, in some senses it's no different from the good old days when God was doing exactly the same thing. Yet it seems acceptable that God should know our private affairs but not the government.

While government may be stepping in where official religions and church attendance are in decline, perhaps a more striking phenomenon is how religious beliefs and behaviors are persisting or reinventing themselves in different guises. Findings in cognitive science suggest that inclinations toward religious thinking will keep cropping up from below, even if organized religion is removed, or a new one or some other replacement is imposed from above. This process has been around for a long time. We learned from the Bible that Moses, for example, had to work hard to abolish the persistent folk religions among the people of Israel as they struggled to recognize the one and only true God. Try as Moses and others might, the Israelites continued to cling on to traditional beliefs in local spirits and lesser gods, and to worship idols. And they did so despite significant dangers that lay in store for such blasphemy. The compulsion, it seems, can be strong.

NEW GODS: NEW CULTS FOR NEW PROBLEMS

In some parts of the world, religion is declining. In others it is increasing. But one thing we can count on is that it is always changing. One driver of change comes from people responding to crises, when their pre-existing beliefs or practices no longer meet the demands of the day. An interesting example of this phenomenon is the emergence of cults in drug-torn Mexico. In recent years, Mexico has seen a remarkable increase in the worship of an unlikely deity, *La Santa Muerte*—Holy Death. *La Santa Muerte* is usually depicted as a skeleton, dressed in long robes and carrying a scythe. At first glance the saint appears similar to the grim reaper, and reminiscent of the figures celebrated in Mexico's Day of the Dead, when the departed come back to feast with their relatives still on Earth. But *La Santa Muerte* is a different beast entirely. A figure of death itself, this saint has nevertheless become attractive because she is the "guardian of the most defenseless and worst of sinners." She has, therefore, become extremely popular among drug traffickers and others involved in the drug trade, as well as regular law-abiding citizens who have been caught up in the violence and danger. Although the Mexican Interior Ministry has clamped down on worship of *La Santa Muerte*, the movement is on the march and is now the fastest growing cult in Mexico. The Interior Ministry itself acknowledges that it "can now be found all over Mexico, on street corners and in the homes of the poor," and it is spreading into Central America and the United States.[11]

La Santa Muerte is one of several supernatural figures that have risen to prominence in recent years in Mexico, including Afro-Cuban deities, a mythical bandit from Northern Mexico called Jesus Malverde (the "outlaw narco-saint"), and saints from the Bible that have morphed from bringing salvation to bringing success. There has also been a rise in the popularity of St. Jude Thaddeus, the patron saint of desperate causes. Followers of St. Jude have started a trend of monthly masses in which "worshippers crawl to the statue of the saint on their knees, praying for help, protection and survival."[12]

So what exactly is the crisis that is argued to underlie these revisionist religious movements? As veteran journalist Alma Guillermoprieto found, people have begun turning to these various supernatural beings as Mexico "has been overwhelmed by every possible difficulty—drought, an outbreak of swine flu followed closely by the collapse of tourism, the depletion of the reserves of oil that are the main export, an economic meltdown, and above all, the wretched gift of the drug trade and its highly publicized and gruesome violence." In recent years, the fight against drugs became a bloody war fought by the army as well as the police, with tens of thousands of casualties on both sides. José Luis González, an expert on popular religions from Mexico's National School of Anthropology and History, explained that "the emotional pressures, the tensions of living in a time of crisis lead people to look for symbolic figures that can help them face danger."[13]

As bizarre as the new cult of *La Santa Muerta* is, it is striking in the light of the supernatural punishment theory. First, there is an adaptive aspect to it. A new religious practice has emerged to help people deal with a novel challenge, offering security and help to lawless parts of society that cannot rely on the help of government or moralizing gods. Second, *La Santa Muerte* does not just stand around looking pretty. She is the purveyor of supernatural rewards and punishments, "who heals, protects, and delivers devotees to their destinations in the afterlife." *La Santa Muerte* "will grant your prayers—but only in exchange for payment, and that payment must be proportional to the size of the miracle requested, and the punishment for not meeting one's debt to her is terrible."[14]

OLD GODS: RELIGION NEVER DIES

China is officially a nonreligious state, and there have been serious attempts to remove religion from society altogether. After Mao's revolution China outlawed religion, and today "the [communist] party's official stance is that religion will die out under socialism." But despite the iron fist of Chinese communist rule for several decades, there has been "a resurgence of religious or quasi-religious activity across China that—notwithstanding occasional crackdowns—is transforming the social and political landscape of many parts of the countryside." The revival has centered around traditional folk religions and veneration of ancestors, as well as the familiar religions of Buddhism, Taoism, Islam, and Christianity. Like with Moses and Mexico, grassroots religion keeps sprouting up despite significant efforts to suppress it. In China, the communist party is being forced to adapt in the face of facts on the ground. And although ideologically opposed to religion, they seem to be coming around to the idea that it can be useful. In many rural areas communities are having to deal with debts and social problems with little government help, so they often turn to local temples which offer a source of financial support for things like schools and community infrastructure.[15]

While the communist party does at least recognize five major religions—Buddhism (with an estimated 100 million adherents), Islam (20 million), Taoism (unknown numbers), Protestantism (16 million), and Catholicism (5 million)—by contrast, the government sees "folk religion as superstition, the public practice of which is illegal." It survives with unofficial tacit approval from local officials and police, who understand the lack of alternatives (and may also be looked after in kind). They cannot be shut down because of the sheer numbers of people involved and the persistence of the desire to practice them, plus the awkward economic reality that they are important sources of support for social welfare, local community projects, and sometimes also for local authorities. An article in *The Economist* reported that "in the countryside, party secretaries routinely take part in religious ceremonies," and temple chiefs are often party bosses as well. As inconvenient as they are for the communist ideal, "they generate prosperity for the local economy and income for the local government." Even the government itself is not immune to the recurrence of religion, "within its own ranks, the party knows that some members practice religion even though this is against the party's rules."[16]

In the spirit of nationalism, the Chinese government has recently been making efforts to revive traditional Chinese culture, and traditional religion has to some extent been part of the revival. At times it has been seen as playing a positive role and being important to China's vision of a harmonious society. Evidently, as with all the great national experiments to exorcise religion from society, from France to Albania to the Soviet Union, it has failed. However harshly or consistently it is put down, religion grows back.

Throughout history, wherever it was crushed or seemed to have disappeared, religion has sprung up again or evolved into new forms. There is no reason to think that the same won't happen again in the current and future context of secularization and the decline of religion, at least in the West. In its place we see not only CCTV cameras and remote surveillance, but also expanding interest in new age religions, the rise of cults, the resurgence of fundamentalism, the renaissance of folk religions, and the difficult to measure but widespread practice of maintaining private religious beliefs in isolation from any official church—"fuzzy fidelity," as religion scholar Ingrid Storm called it. The replacement for God, therefore, may simply be more gods.[17]

We began the chapter talking as though the spread of secularism and the demise of religion is self-evident and inevitable. But when we turn our attention away from western societies and look at the globe as a whole, we see that the role of religion in human society is not a quirk of the past, but rather an entrenched feature of the modern world. A great majority of the world's population are religious, and in many regions the number of believers and the levels of belief are increasing, not decreasing. In a 2006 *Foreign Policy* article entitled "Why God is Winning," Timothy Samuel Shah of the Council on Foreign Relations and Monica Duffy Toft, then of the Kennedy School of Government at Harvard University, reviewed the evidence on how religion has become an increasingly important issue in domestic and international politics, spanning elections, public policy, international relations, and war. The authors pointed to religion's "global politicization," a process that can be traced to key events of the late 1970s—the election of an evangelical, Jimmy Carter, to the US presidency, Indira Ghandhi's suspension of democracy in India in response to the threat of the Hindu nationalists, the ascendancy of Pope John Paul II, the Islamic Revolution in Iran, and the Sunni seizure of the Great Mosque at Mecca.[18]

The trend continued with the end of the Cold War, as the officially atheistic Soviet Union collapsed to reveal many religiously backed independence movements around Europe and Asia. Such trends have affected the United States as much as anywhere else. There has been an increasing influence of the religious right in US presidential elections (40 percent of the George W. Bush vote in 2004), as well as in lobbying groups for domestic and foreign policy. And of course, 9/11 helped to bring religion to the top of the political agenda not just in the United States but around the world. The "war on terror" may have been a misleading term of propaganda, but it is nevertheless clear that US strategy over the last decade has primarily targeted extremist Islamic movements. As security expert William Rosenau noted in 2007, "combating al-Qaida has become the central organizing principle of U.S. national security policy." Even as the operations in Afghanistan and

Iraq came to an end, religion continues to be an important aspect of new threats emerging in Syria, Iraq, Iran, Israel-Palestine, Pakistan, India, and Nigeria, among many others.[19]

Globalization itself is likely to exacerbate these trends. As democratic and secular governance has spread, along with it there has been a striking decline of liberal religions and a resurgence of fundamentalism—what French Islamic scholar Gilles Kepel calls "the revenge of God." The spread of fundamentalist and proselytizing forms of religion is not, of course, limited to Islam; there are powerful such efforts around the world by evangelical, Pentecostal, and charismatic Christian groups as well. It is sometimes thought that following the Treaty of Westphalia and the Enlightenment, politics had escaped the chains of religion (or perhaps the other way around). But Kepel argues that the considerable religious revivalism in Islam, Catholicism, North American Protestantism, and Judaism witnessed in recent decades is a reflection of widening and deepening discontent with the modern world and its politics. These new religious movements are aimed at "recovering a sacred foundation for the organisation of society." He argues in particular that the aim in Islam is no longer to "modernize Islam but to Islamize modernity."[20]

In other respects, the striking feature is not so much that religions have strengthened, but that religions have stubbornly persisted even though the world around them has changed. In the United States and many other countries, including several in the West, beliefs go from strength to strength. One recent analysis of survey data found that despite large shifts in its practice around the world, "religion remains a powerful conservative force in people's behavior." It also continues to be reflected in institutions, law, and culture. In particular, religious injunctions to enforce cooperation and deter defectors still have considerable currency in modern legal and political discourse, such as swearing on the Bible to tell the truth in court, the appeal to religious duty at home and abroad, and warnings of the dangers of evil forces in international relations.[21]

But increases are evident as well. One study found that although religious *adherence* in general has not changed a great deal in the United

States since the 1960s, critical *aspects* of religiosity are rising in prominence, with increases in the proportion of Americans who strongly agreed with the following: that there was no doubt God existed, that they will be called before God to answer for their sins on judgment day, that God performs miracles in today's world, that prayer was an important part of their daily life, and that clear guidelines distinguishing good from evil apply to everyone. Most remarkably, these increases were observable across a variety of denominations including evangelical, mainline, African American, and Catholic churches, and even—yes, really—atheists! Fully 34 percent of self-declared atheists, for example, expected to be called before God on judgment day to answer for their sins. This, of course, hints at important differences between "Christian" atheists and atheists in the communities or societies of other religions—all of us are influenced by our cultural surroundings whether we belong to a particular religion or not.[22]

Broader demographic trends are perhaps the simplest but most striking of all: They reveal an increasing prominence of religious adherence across the globe as a whole. Research by Pippa Norris and Ronald Inglehart found that in the century between 1900 and 2000 there was a significant increase in the *proportion* of people declaring adherence to Catholicism, Hinduism and, most dramatically, Protestantism and Islam. These four religions accounted for 50 percent of the world's population in 1900, but that rose to 64 percent in 2000. Birth rates are likely to fuel this trend further. Between 1997 and 2015, the average population growth among the most religious societies is estimated to be 1.5 percent, while among the most secular societies it is only 0.2 percent, and falling. Higher birth rates may or may not be *caused* by religiosity. The point here is simply that religious societies are growing faster than secular ones.[23]

Across the world, religion is increasing in vitality, numbers of adherents, and influence on society and politics. Partly despite and partly because of globalization, we are faced with the prospect of expanding religious groups coming in to ever-closer competition with each other over fundamental values, dwindling resources, and global ambitions. To many people, rightly or wrongly, these trends seem to point to a coming clash of civilizations. Whether true or just perceived that way, it is more important than ever that we seek to understand the role of religion in cooperation and conflict. Only then will we know what to do about it.[24]

SHOULD WE GIVE UP GOD?

The eyes of God, so important to societies of the past, may be misting over as official religions decline, at least in many parts of the western world. But even here we have seen that they are being replaced by other forms of surveillance that watch us instead: police, big government, Internet companies, CCTV, credit card agencies and advertising giants, the GPS in your phone, and so on. Someone—or something—out there knows exactly what you have been doing today. We know all the reasons why this may be a bad thing. But perhaps it can be a good thing, too. Among other consequences, it may have the effect of deterring selfish behavior and encouraging cooperative behavior—just like supernatural punishment.

But does it work as *well* as God? There are, in fact, reasons to doubt it (Table 8.1; here I write "God" for simplicity but the logic stands for other supernatural agents as well, even if their powers are somewhat more limited). For a start, religious believers typically have an intimate relationship with God. God loves them and they love God. People don't love big government and surveillance cameras (and they

Table 8.1 A comparison of the effectiveness of secular and supernatural policing.

	God	Law
Respected	Yes	No
Surveillance	Complete	Incomplete
Punishment	Infinite	Limited
Effectiveness	Perfect	Imperfect
Naturalness of concept	High	Low

don't love us). Perhaps Big Brothers only work smoothly and effectively if there is a reciprocal relationship of mutual respect. Second, we saw in Chapter 4 that a range of features of *supernatural* agents are better able to solve many of the game theoretical problems of cooperation (as long as people believe in them, of course). God is always going to be better than CCTV, because He is everywhere at all times. He misses nothing. We would need a lot more technology and invasiveness to come even close to the level of monitoring that God provides. And even then, we'd still only be able to monitor people's behavior. Motives, plans, and intentions, at least, would still remain private. But God can see those too.

Our powers of punishment are limited as well, and they are becoming ever more tame. In the past, people were dissuaded from crime by the prospect of dungeons, torture, and death. Today, people are dissuaded from crime—at least in most western countries—only by fines or time in prison. It may be more humane, but its less of a deterrent. Still, whatever the differences between medieval and modern deterrents, all of these punishments are pathetic compared to an eternity in hell. For a true believer, that is a deterrent indeed. Even for an agnostic or an atheist, who may think God is vanishingly unlikely, the danger of rejecting Him remains significant because the consequences of being wrong are so colossal. A one in a billion chance of God being real, multiplied by an eternity in Hell, makes for a pretty lousy expected utility calculation. This, of course, was Pascal's wager—and again it is the negative side of the equation that seems to loom especially large. God's unrivaled powers of punishment may remain a much more powerful deterrent than anything secular institutions can come up with—not only for believers but for skeptics too.[25]

Secular institutions are also necessarily imperfect. They may, for example, be understaffed, overcommitted, inefficient, inattentive, corrupt, slow, or outwitted, none of which are problems for God. Or they may even go AWOL. Cognitive scientist Steven Pinker grew up in Montreal and recounts an episode when the city's police once went on strike in 1969. The citizens waited with baited breath to see what would happen. Would the city descend into chaos? Or would life just go on as normal, revealing the wasted resources our societies invest in policing people who are perfectly able to behave on their own? People didn't have to wait long to find out. Within minutes of the strike beginning, there was a spate of burglaries, bank robberies, murders, rioting, and looting. The experiment that no one had ever wanted to carry out coughed up its results in short order. Police may be necessary, but they are not infallible. God, of course, does not go on strike.[26]

Finally, whatever the effectiveness of secular monitoring and punishment, people may not find it convincing or concerning. We may not even have the cognitive wherewithal to gauge it properly. We did not evolve to worry about security cameras and phone tapping. We evolved to worry about other human beings and supernatural agents. God may therefore have a competitive edge simply because we have a pre-existing cognitive template into which he fits. Surveillance cameras rely on us conceptualizing and remembering, in the heat of the moment, that there may be a plastic box somewhere nearby nailed to a wall. Moreover, as sociologists Paul Robinson and John Darley showed in a large sample study of criminal behavior, potential offenders tend not to know what the laws are, to assume the police are unlikely to detect their offenses, and to believe that even if they did they would be unlikely to be punished. In situations of high emotion or fast-moving events, our brains do not allow us the luxury to think all these evolutionarily novel factors through. We do not, however, have any of these problems with supernatural agents. As the cognitive scientists have discovered, they are *natural*, automatic aspects of our cognition, always present, powerful, and persuasive. If we want good social programs to deter selfishness and encourage people to cooperate, we should make sure they are compatible with the preinstalled hardware in our heads.[27]

CAN WE GIVE UP GOD?

There is another reason why the widely vaunted decline in religion in western countries may be misleading. Even where attendance at churches is dwindling and fewer people are officially affiliated to a particular religion, the idea that people have simply dispensed with their religious beliefs cannot be concluded from these facts. First of all, we've already seen that many people continue to hold private beliefs about God, and about heaven and hell, even if they don't go to religious services or declare allegiance to any religion. Second, human brains won't just suddenly stop believing in supernatural agents, or in the supernatural consequences of our actions, because they can't. The human brain is wired to adopt supernatural concepts whether we belong to a church or not. We can aspire to atheism, rationalize our behavior, and learn science, but we cannot switch off the cognitive mechanisms whose job it is to perceive agency, intentionality, and purpose in the world. As Jesse Bering put it, removing supernatural agency from human minds "would require a neurosurgeon, not a science teacher." Supernatural beliefs, therefore, cannot be in decline any more than any other ingrained characteristic of human nature. All that is happening is that cultural vehicles for these beliefs are manifesting themselves in new and different ways, not least in our unquenched appetite for science fiction, fantasy, ghost stories, conspiracy theories, extraterrestrial intelligence, rituals, mythical figures, superstitions, spiritual experiences and fulfillment, and the proliferation of cults and new age religions—and, of course, the persistence of private religious beliefs. People's brains haven't changed. In our modern interconnected society we just find new ways to vent our innate beliefs in supernatural agency and causation. We might not like to admit this, of course.[28]

A final reason why religious and supernatural beliefs are likely to persist even against the odds is a counterintuitive one: standards of evidence. Bruce Hood points out that "scientists reject beliefs until they are *proven* beyond reasonable doubt. In contrast, supernaturalists accept beliefs until they are *disproven* beyond reasonable doubt. The problem is that it is impossible to disprove anything." This represents a fundamental asymmetry in the practice of science and the persistence of supernatural beliefs. All else being equal, in the minds of laypeople the latter will always win. This is exacerbated by a further asymmetry in the *salience* of natural and supernatural phenomena in the first place. Supernatural phenomena may simply represent more intuitive explanations to human beings, whatever the implications for proving them. As E. O. Wilson observed, "the human mind evolved to believe in the gods. It did not evolve to believe in biology." New work in cognitive science strongly emphasizes this problem. Robert McCauley has described in detail how not only is religious thinking natural, automatic, and easy, but science is the opposite. Learning religion is *part* of human nature. Learning science is a battle *against* human nature. Science therefore has an uphill struggle if it is to knock superstition, let alone religion, off its perch. As long as we have human brains, we will have supernatural beliefs and practices, and wherever these are shared among communities of people, we will have religion. We might therefore paraphrase the pulpit cry about old Europe's unbreakable line of Kings: "God is dead" (says Nietzsche), "long live God!" (says the cognitive scientist).[29]

CONCLUSIONS

The New Atheists—notably Jerry Coyne, Richard Dawkins, Daniel Dennett, Sam Harris, and Christopher Hitchens—argue that religion is bad. It is bad for people, bad for children, bad for society, and bad for the world. Whatever else it does, they stress, it fosters ignorance, hostility to science, sexual discrimination, violations of human rights, justifications for extremist projects or policies, racism, conflict, and war. The way out of the pit, they conclude, is to get rid of God. These are serious charges, and we can all agree that any real or perceived role of religion in such negative outcomes must be identified, researched, and dealt with. But good luck to them with their solution. God cannot die, we have learned, so long as human beings have human brains. But the point I want to end with is not that their mission is flawed because it is impossible, but that the attempt would be counterproductive.

I have argued that religion has a special power to temper self-interest and promote cooperation. If this is the case, then we may not want to get rid of it. Of course, there may be ways to adapt it so that it works better in a globalized, technologically sophisticated society, and indeed this is what theologians and religious leaders are busy doing all the time. But we might need to keep the foundations intact. Sam Harris claims that "all that is good in religion can be had elsewhere." By which he means other forms of ethical and spiritual practices or beliefs that do not have to rely on "things we manifestly do not know." I am not so sure. While religion is not the *only* means by which humans can find ways to cooperate and flourish, it is an unusually good one. And it might be effective precisely *because* it is supernatural and not material—things we manifestly cannot know can be mightily inspiring. Its very supernatural character allows religion to be a more formidable, infallible, and ever-present deterrent than any secular system could be. As we have seen, supernatural punishment offers an uncannily powerful solution to many of the game theoretical problems of cooperation. It may be only one way to keep society together, but its remarkable recurrence in all known human societies across the world and throughout history—what biologists would call "convergent evolution"—seems to attest to its power in steering people's behavior and its effectiveness in achieving results. It may even have been the game changer in the rise of human civilizations, and increasing the scope of what humanity has been able to accomplish since. Indeed, it might yet increase the scope of what humanity is able to achieve—or avoid—in the future.[30]

Religion may or may not be the best way to engineer social cooperation. Like most things, it has pros and cons. And like most things in nature, natural selection (and government) has to work with what it's got—notably human beings with brains susceptible to cause-and-effect reasoning, mind-body dualism, hyperactive agency detection, beliefs in a Just World, and above all, a concern for the supernatural consequences of our actions. But the more immediate point is this. Shouldn't we be investing time and energy into examining exactly whether, how, when, and why religion *does* improve cooperation? (And when it fails). To assume that this complicated old machine we have found in our evolutionary garage is of no use, and to fling it onto the scrapheap of history seems rather hasty. We might need it later. Or we might at least want to take it apart to explore how it works if we are going to build an effective replacement. And this is a scientific call for action, not a religious one. As an editorial in the *New Scientist* noted, "to borrow from a popular biblical saying, humankind cannot live by rational thought alone. To want to cleanse society of religion before understanding its evolutionary roots and purpose seems strangely unscientific."[31]

If the expectation of supernatural punishment for selfish behavior is not an accidental byproduct of evolution, but an adaptive tool favored by natural selection, then rather than trying to purge it from our minds it might give us a reason to have faith in our instincts—even our religious instincts. People evolved to believe that our behavior is beholden to supernatural consequences as well as whatever potential but unreliable secular consequences are available, and maybe we should embrace these beliefs rather than try to eradicate them. However much we may dislike the idea of guardian angels watching over our own shoulders, we like to think that other people have them. Belonging to a religion is a way of telling whether other people are at least ostensibly playing by the same rulebook as you. That knowledge increases the prospects for trust, cooperation, and productivity with other members of the group, while simultaneously providing a means of avoiding any free-riders poised to exploit you. The New Atheists' mission of creating a godless world is an untested experiment that is likely to have negative as well as positive consequences. But we have little idea yet what any of them might be. Are we playing with fire?

CHAPTER 9

GOD KNOWS

Amazing grace, how sweet the sound,
That saved a wretch like me.
I once was lost, but now am found,
Was blind, but now I see.

T'was grace that taught my heart to fear,
And grace my fears relieved.
How precious did that grace appear
The hour I first believed.

Through many dangers, toils, and snares,
I have already come;
'Tis grace that brought me safe thus far,
And grace will lead me home.

—John Newton (1779)[1]

Roughly two hundred and twenty thousand people died in the 2010 Haiti earthquake. But one, Evans Monsigrace, was found alive after twenty-seven days buried under the rubble, "confounding doctors and defying medical logic." A father of two, Monsigrace had been a vendor selling rice and cooking oil at the La Saline marketplace in Port-au-Prince. He had just sold the last bag of rice when his world literally fell apart. Concrete showered down, trapping and burying him. Yet somehow he survived the initial destruction, as well as almost a month of internment that followed. As a journalist later reported from his hospital bedside in Tampa, Florida, "for the last 10 weeks, Evans Monsigrace has struggled to understand how and why he is still alive. So remarkable is his survival, that at times it has been easier for him to think that he must in fact be dead."[2]

The Haitian government had ended the search and rescue effort on January 23, thirteen days after the earthquake had struck. But a few survivors continued to be found. Sixteen-year-old Daline Etienne was pulled out from under her collapsed house fifteen days after the quake, without major injuries. The rescue was "deemed a miracle." As time went on, the miracles only became greater. But Evans Monsigrace was the last and greatest miracle. Monsigrace's mother said, "I thought he was dead, but God kept him from dying." Evans, recovering in intensive care, felt the same way: "I was resigned to death. But God gave me life. The fact that I'm alive today isn't because of me, it's because of the grace of God. It's a miracle, I can't explain it." He recalled that as masonry fell around him, a big slab of concrete appeared to be "pulled back" at the last moment, which he attributed to God, or one of Haiti's voodoo snake spirits. And then he had to survive his entombment. Early reports suggested that someone must have brought him water, but it turns out they did not. He had been completely alone and buried deep underground. Pinned down by concrete, he saw nobody for a month. It seems he survived by sipping a trickle of sewage water seeping past. He lost sixty pounds and suffered severe dehydration but, amazingly for someone who has gone for so long without potable water, his kidney function was good. Whether one takes a supernatural or a medical perspective, as Dr. David Smith of the Tampa hospital remarked: "He calls himself a miracle? He's right."[3]

Was Evans Monsigrace a victim of supernatural punishment, or the beneficiary of supernatural aid? The same event can be interpreted in radically different ways. Stunned by the scale and randomness of the disaster, people had asked, How could God have caused or allowed such an event to happen? And what had Haiti done to deserve it? But the same line of thinking also raises the exact opposite question: How could God have caused or allowed certain people to *survive*? And what had they done to deserve *that*? Why did this person die while that one escaped? Did God work to protect people amidst the carnage? Against the backdrop of utter and indiscriminate devastation, the stories of survival against the odds might just as well point to divine providence instead of divine vengeance. Certainly, it is possible to see it this way, and to some this seemed almost the only possible explanation given the ubiquity of death and destruction.

The words of John Newton's *Amazing Grace* in the opening epigraph remind us that the notion of God, at least as imagined in modern Christianity, is not of course only about divine retribution, wrath, and punishment. For many people, God represents rewards rather than punishments, or neither punishment nor rewards, but love and grace. John Newton came up with the words for *Amazing Grace* after surviving a harrowing storm at sea. Even when times are bad, or indeed precisely because they are bad, we may be especially cognizant of the benevolence of supernatural agency. That Evans Monsigrace survived the Haiti earthquake seems almost as remarkable as the multitude who did not.

While recognizing, therefore, a role for supernatural benevolence, this book has focused on the special power of the other side of the coin—punishment. The punishment that, although we all might wish we could be rid of it, is the crowbar that can lever cooperation into motion, maintain it in the face of adversity, extend it to large groups of anonymous strangers, and generate greater social goods than would otherwise be possible. We often have to accept things we don't like in to order to get what we want. We don't like to exercise, but we enjoy the benefits. We don't like paying taxes, but we live in a better society for it. We don't like to clean up, but we want a tidy place to live. In many aspects of life, we must endure negative experiences to attain positive goals. No pain, no gain. And so it is with cooperation. We all want the rewards that cooperation offers, but in a world of self-interested individuals, and the ever present threat of exploitation by free-riders, we need some mechanism of enforcement. Sometimes, we may be able to simply ignore the free-riders and choose to interact with other cooperators instead. But in large societies of vague acquaintances and anonymous strangers that we may rarely or never meet again, we cannot be so choosy. We rely instead on institutions that can deter and punish those who would take advantage of others. This book has examined the role of one such institution, but perhaps the most powerful one that humans have ever enjoyed: religion.

RELIGION IN THE AGE OF BIOLOGY

If people weren't already interested in religion in recent years, they are now. The rising influence of Christian conservatives; clashes over constitutional laws surrounding abortion, evolution, and cloning; 9/11 and the resurgence of fundamentalism around the world; conflicts in Afghanistan, India, Iraq, Nigeria, Pakistan, and Syria; Iran's development of nuclear technology; the ongoing Israeli-Palestinian conflict; and the Arab Spring—all put religion center stage in the global news and in our local lives. What do we do about Islamic terrorism? How much do religious lobbies and voters influence US domestic and foreign policy? Will your child marry an atheist? Or an extremist? What are your children going to learn at school? Will American laboratories be allowed to clone the bone marrow that might save your life? And, not least, what is the legacy of your personal life? What good have you done on this Earth, and what bad have you done? Will it matter?

At the turn of the twentieth century, people were still just learning about Darwin. At the turn of the twenty-first century, bolstered by rapid advances in genetics, molecular biology, and evolutionary science, Darwin's theory has emerged triumphant as the single best theory to explain the entire 3.5 billion year history of life on Earth and all its diversity, including human genetics, physiology, brains, and behavior—as well as many aspects of culture. Some are calling it the "age of biology." There has been no time like the present to examine the evolutionary origins and consequences of religion. Just as we have asked of other human behavioral traits, we can ask, What is religion *for*? What does it make people do? How might it help us? Religion is a phenomenon of stunning complexity and diversity. But one thing that ties it together is a fundamental characteristic of all religions, and indeed, not coincidentally, of human nature—we live our lives as if surrounded by watchful supernatural agents with formidable powers of payback. For our genes' own sake our selfishness must be kept in check, and our cooperativeness must be unfurled. A guiding hand to help us do this is likely to have been highly adaptive in human evolution, as it often still is today. While religion may sometimes fan the flames of conflict or discourage freedom of thought, these costs may be insignificant compared to the less visible but far more pervasive benefits of supernatural beliefs in the everyday lives of billions of believers.

As I have stressed, supernatural punishment is not the be-all and end-all of religion, nor of cooperation. It is a cog in the machine. I do think, however, that it is a vital cog. Without it, religion does not work. What it does is to put in motion all of the other elements—a critical addition that turns human groups from a heap of promising components into well-functioning and robust societies. As psychologist Matt Rossano argues, religion may have only been one factor among many, but a decisive one in tipping the balance toward cooperation. This book has argued that supernatural punishment is a key driver, playing a powerful role in achieving and sustaining cooperation—and it does so among modern, ancient, and indigenous societies, as well as holding up to scrutiny in controlled laboratory experiments. Supernatural punishment is not only widespread, it is also a significantly more potent weapon than we might have thought, because human beings and human brains are particularly susceptible to *negative* events, and because *supernatural* punishment is inherently more powerful than secular punishment. Without supernatural punishment, therefore, other factors setting the scene for the evolution of large-scale cooperation may have been in vain.[4]

By the turn of the twenty-second century, many people seem to think, religion will be dead. Religions are on their way out in many western countries and the trend may simply continue as secular democracies spread. But they're wrong. Religion will be alive and well. Liberal European churches may be going out of fashion, but the religious beliefs and behaviors of human beings are here to stay. We can no more get rid of religion than we can get rid of our brains or our appendix. It's part of our biology. An evolutionary perspective suggests how and why religion evolved, when it tends to help or hurt us, and how we can channel it to benefit ourselves, our societies, and the wider world.

EXTENDING THE REACH
OF SUPERNATURAL PUNISHMENT

The supernatural punishment theory sheds new light on both the evolution of cooperation and the evolution of religion. Whatever the variation in beliefs about supernatural punishment across societies, religions, churches, and individuals, it has evidently played an important role in the origins and evolution of religion. If it is so far-reaching and so important in human psychology and behavior, then we might also expect it to help explain some other features of social life that have until now remained obscure. While a number of candidates suggest themselves, two examples in particular that we have touched on in the book are Just World beliefs and the origins of ritual.

PUNISHMENT AS A DRIVER
OF JUST WORLD BELIEFS

Despite being a well-documented and intensively studied phenomenon, the origins and utility of Just World beliefs remain unclear. *Why* do people think this way? Andre and Velasquez noted that "neither science nor psychology has satisfactorily answered the question of why the need to view the world as just exerts such a powerful influence on human behavior and the human psyche." An explanation is offered, however, by the supernatural punishment theory. The reason we gravitate to Just World beliefs, and why it has such a powerful impact on human psychology and behavior, is because the expectation of supernatural reward and punishment is an adaptive feature of human brains. As I have argued, it provided significant fitness benefits in suppressing self-interest and promoting cooperation in our evolutionary past. Just World beliefs are studied primarily as people's expectations about how *other* people's conduct results in good or bad fortunes. But we are hardly immune to this effect ourselves—we also tend to think *we* will get what we deserve. And it is the influence of this expectation on our own behavior, rather than on the behavior of others, which appears to hold the key. The key role of genetic self-interest in evolutionary theory may have provided the right lens to bring this point into focus.

Our cognitive mechanisms, behavioral dispositions, and cultural traits have become organized around the idea that our actions bring not only material but also supernatural consequences—and it is precisely the supernatural ones that may be more reliable, pervasive, and powerful. Just World beliefs are not a random quirk of human behavior, but seem to represent an adaptive moderator of social behavior that we expect to apply to others as well as ourselves.[5]

PUNISHMENT AS A DRIVER OF RITUALS

A second major area onto which the supernatural punishment theory may shed new light is ritual. In research on religion in general, ritual is big. In *evolutionary* research on religion, ritual is even bigger. Scholars increasingly view religious ritual as a natural extension of the logic of ritualistic behavioral patterns found in other animals, which function as honest or hard-to-fake signals of commitment that bind individuals together. Notably, evolutionary *psychologists* who work on religion have focused on cognitive factors and beliefs. By contrast, evolutionary *anthropologists* who work on religion have focused on behavioral factors and ritual. That is no surprise, but it does suggest a disciplinary bias to see one perspective as intrinsically more important than the other. At the moment, people on each side are arguing over which is *the* key thing about religion. I was trained as an evolutionary biologist, and therefore have no particular stake in the relative role of psychology versus behavior. Both are clearly important. Indeed, I suggest that both beliefs *and* rituals stem from the same underlying principle: anticipation of supernatural consequences.

Rituals may have benefits that arise from their mere performance alone, such as demonstrating commitment and unity. But if that were their only function, rituals could be arbitrary. We could, say, carry sixty bricks around the block every other Tuesday. But people do not perform arbitrary rituals. They perform rituals that are thought to *do* something—to have consequences in the supernatural realm as well as in the physical one. Rituals are designed to invite supernatural benevolence (such as rain dances or communal prayer), or to avoid supernatural vengeance (such as sacrificial offerings or libations). If rituals had no consequences, no one would perform them. Even if the major function of ritual *is* mere signaling to others, these benefits would be all the more effectively achieved through rituals that have consequences rather than rituals that have no consequences. The former should thus outcompete the latter in cultural selection. A belief in supernatural consequences, therefore, is critical. It is the keystone without which collective rituals lose their salience.[6]

Research into the cognitive underpinnings of ritual behavior supports this idea. Humans appear to find rituals compelling precisely because they are a way of controlling outcomes in the environment. From the psychiatric extreme of obsessive compulsive disorder to the everyday superstitious acts of mentally healthy adults, rituals appear to be designed to repeat behaviors that are consciously or subconsciously believed to bring about good outcomes and avoid bad ones. And once again we find a striking asymmetry: Bad outcomes matter more. Psychologists find that individual ritual behavior tends to be about avoiding possible dangers rather than attracting possible benefits. The *thoughts* underlying ritual behavior "revolve around a limited number of [mainly negative] themes, such as contagion and contamination, aggression, and safety from intrusion," while the *behaviors* themselves also focus on negative things, such as "washing, cleansing, ordering and securing one's environments, or avoiding particular places." The reason such beliefs become obsessive is because "there can never be positive evidence that a potential danger has been eliminated." Avoiding supernatural punishment, in other words, is not just a compulsive concern, but a never-ending one. Religious rituals incessantly reinforce the knowledge and practice of acceptable behavior and the discouragement of unacceptable behavior, but the prominent theme is steering one away from the wrath of the gods. As religion scholar Joseph Bulbulia put it to me, they are "a kind of boot camp for tamping down impulse. Most rituals are about denial." Rituals may or may not work in attaining dividends, but they certainly must be done to avoid disaster. Anthropologists may dislike this imperious claim, but it seems to me that even rituals are mainly about the avoidance of supernatural punishment (or at least avoiding the withholding of supernatural benevolence if the rituals fail to be carried out).[7]

THE GOOD OF WRATH

We have been to the moon, split the atom, decoded our genome, and mapped the brain. Yet science has made precious little headway into one of the biggest scientific puzzles of all time. Why do people believe in God?—or in *some* god, gods, spirits, or other supernatural beings or agency? Why, across the globe, over all of recorded history, and in every indigenous society ever known, did humans so reliably come up with religion? And why is it always accompanied by a seemingly rather specific belief in supernatural punishment?

Religious observance consumes precious time, effort, resources, and opportunity costs. Believing in supernatural punishment is even worse, because it deters you from self-interested pursuits that might bring great benefits. So why didn't evolution crush such costly beliefs in invisible agents under the unforgiving millstone of natural selection? Even Richard Dawkins, who generally argues that religion is a malicious meme, a parasite on human brains, concedes that "because Darwinian natural selection abhors waste, any ubiquitous feature of a species—such as religion—must have conferred some advantage or it wouldn't have survived."[8]

One possibility is that a belief that God, or *something*, is watching over our actions, even our thoughts, and poised to reward or punish us, assists us in some way. In particular, it might act as a guardian angel, not to protect us from the dangers of the environment or other people, but to protect us from ourselves—making us good in a world where being bad can be costly to Darwinian fitness. Selfishness and egoism were good strategies for 3.5 billion years of evolution, but they became dangerous in the socially transparent world of cognitively sophisticated human beings. The prospect of retaliation and reputational damage—even from absent third parties—became a significant concern. Restraint may thus have started to pay greater dividends where selfishness had presided in the past.

Since even atheists harbor a conscious or subconscious anticipation of supernatural consequences of their actions (as we found in Chapter 5), religion does not have a monopoly on supernatural punishment promoting certain behaviors and preventing others. We even reinvent supernatural punishment when it helps to promote cooperation, scolding our children for doing something naughty in the name of what Santa Claus or their father would think—even if they're miles away or at the North Pole. We can also find ourselves motivated by the efforts and sacrifices of long-dead forebears, whether the founding fathers of a nation, historical idols, or our own ancestors. What would *they* think of you and your actions today? It's illogical because they can't do anything about it, yet it is commonly invoked and exerts a real pull on our thinking and behavior. But among all methods of invoking a fear of consequences for our actions, religion has exploited this feature of human nature most effectively, anchoring a nagging sense of payback to the wiring of a mammalian brain sensitized to threats and dangers, and sometimes ramming the message home with explicit images such as damnation and hell, lorded over by powerful agents that are omniscient, omnipotent, moralizing, infallible, punishing, and yet remarkably alluring.

Religion is complicated. It involves beliefs, rituals, metaphysics, unknowable propositions, history, culture, society, scriptures, interpretation, personal experiences, leadership, followership, doctrine, practicality, peers, death, life, meaning, understanding, and on and on. However, all of these multifaceted concepts revolve around one thing: consequences. If people thought there were absolutely no consequences of religious beliefs and behaviors whatsoever, then they would not pursue them. People adopt certain beliefs, follow certain rituals, and organize religions around them because they believe they have important consequences in the supernatural as well as the material realm. These consequences can be positive or negative. However, what we have learned is that negative consequences are more powerful, more decisive in natural selection, more salient to human psychology, more significant in our neurobiology, and better at suppressing self-interest, deterring free-riders, and facilitating the remarkable collective endeavors that human societies have accomplished.

We could not have got close to the moon, splitting the atom, decoding our genome, or mapping the brain without something so important to human society that we often forget or don't even notice its miraculous effects at all: cooperation. Hard to achieve, hard to sustain, hard to explain. Yet cooperation is vital to our existence. It is the source of our success, indeed what defines us as a species. It is the root of our greatest triumphs and our bloodiest tragedies. It is what makes our societies and what makes our history. And it is what will make our future. No other species can approach the scale and power of human cooperation and the opportunities—and dangers—it generates. While there are many factors that help to create cooperation, supernatural agents appear to have a remarkably golden touch. At least, that seems to be what every society that has ever lived has found out for themselves. But in order to help self-interested and fallible humans get along, the gods have had to be cruel to be kind.

NOTES

CHAPTER 1

1. Stevenson, Robert Louis, *Treasure Island*, Project Gutenberg eBook.
2. Zuckerman, P., *Society Without God: What the Least Religious Nations Can Tell Us About Contentment* (New York: NYU Press, 2008).

CHAPTER 2

1. Baumeister, R. F. et al., "Bad Is Stronger Than Good," *Review of General Psychology* 5, no. 4 (2001): 323–70.
2. "Flawed Dream of Free Cycle Schemes," *The Guardian*, January 5, 2007.
3. Williams, G. C., *Adaptation and Natural Selection* (Princeton, NJ: Princeton University Press, 1966).
4. Nowak, M. A., "Five Rules for the Evolution of Cooperation," *Science* 314, no. 5805 (2006): 1560–63. The fifth mechanism referred to in Nowak's paper is "network reciprocity," where cooperation can arise because individuals are more likely to interact with certain other individuals because of spatial or other constraints, and thus individuals more tightly connected in the network have an interest in establishing cooperation knowing that their mutual interactions are likely to be repeated.
5. Fehr, E., and Gächter, S., "Altruistic Punishment in Humans," *Nature* 415, no. 6868 (2002): 137–40, p. 137; Henrich, J. et al., eds., *Foundations of Human Sociality: Economic Experiments and Ethnographic Evidence from Fifteen Small-Scale Societies* (Oxford: Oxford University Press, 2004).

6. Poundstone, W., *Prisoner's Dilemma: John von Neumann, Game Theory and the Puzzle of the Bomb* (Oxford: Oxford University Press, 1992); Axelrod, R., *The Evolution of Cooperation* (London: Penguin, 1984).

7. Axelrod, *The Evolution of Cooperation*. Robert Axelrod famously explored the point that, if games are repeated many times, both players can be better off by cooperating, since over multiple games the payoffs from cooperation will exceed the payoffs from mutual defection. The problem, however, is how to establish cooperation. Defection is still the best strategy in a one-shot game, or in a limited series of games, since no one wants to be the only one, or the first one, to be a sucker. Something else is needed to get cooperation of the ground.

8. Hardin, G., "The Tragedy of the Commons," *Science* 162, no. 3859 (1968): 1243–48; Ostrom, E., *Governing the Commons: The Evolution of Institutions for Collective Action* (Cambridge: Cambridge University Press, 1990); Ostrom, E., and Nagendra, H., "Insights on Linking Forests, Trees, and People from the Air, on the Ground, and in the Laboratory," *Proceedings of the National Academy of Sciences* 103, no. 51 (2006): 19224–19231.

9. Fehr and Gächter, "Altruistic Punishment in Humans."

10. Fehr, E., and Fischbacher, U., "The Nature of Human Altruism," *Nature* 425, no. 6960 (2003): 785–91.

11. Gürerk, Ö., Irlenbusch, B., and Rockenbach, B., "The Competitive Advantage of Sanctioning Institutions," *Science* 312, no. 5770 (2006): 108–11, p. 108.

12. Burton-Chellew, M. N., and West, S. A., "Pro-Social Preferences Do Not Explain Human Cooperation in Public-Goods Games," *Proceedings of the National Academy of Sciences (PNAS)* 110, no. 1 (2013): 216–21; Burnham, T., and Johnson, D. D. P., "The Biological and Evolutionary Logic of Human Cooperation," *Analyse & Kritik* 27, no. 1 (2005): 113–35; Camerer, C. F., "Experimental, Cultural and Neural Evidence of Deliberate Prosociality," *Trends in Cognitive Sciences* 17, no. 3 (2013); Cronk, L., and Leech, B. L., *Meeting at Grand Central: Understanding the Social and Evolutionary Roots of Cooperation* (Princeton, NJ: Princeton University Press, 2013); Johnson, D. D. P., Stopka,

P., and Knights, S., "The Puzzle of Human Cooperation," *Nature* 421, no. 6926 (2003): 911–12; Sigmund, K., "Punish or Perish? Retaliation and Collaboration Among Humans," *Trends in Ecology & Evolution* 22, no. 11 (2007): 593–600; Ostrom, *Governing the Commons*; Pedersen, E. J., Kurzban, R., and McCullough, M. E., "Do Humans Really Punish Altruistically? A Closer Look," *Proceedings of the Royal Society B: Biological Sciences* 280, no. 1758 (2013); Price, M. E., "Monitoring, Reputation, and 'Greenbeard' Reciprocity in a Shuar Work Team," *Journal of Organizational Behavior* 27, no. 2 (2006): 201–19; Price, M. E., and Johnson, D. D. P., "The Adaptationist Theory of Cooperation in Groups: Evolutionary Predictions for Organizational Cooperation," in *Evolutionary Psychology in the Business Sciences*, ed. Saad, G. (Springer, 2011).

13. Machiavelli, N., *The Prince* (New York: Norton, 1992 [1532]), Chapter XVII.

14. Baumeister et al., "Bad Is Stronger Than Good."

15. Smith, N. K. et al., "May I Have Your Attention, Please: Electrocortical Responses to Positive and Negative Stimuli," *Neuropsychologia* 41, no. 2 (2003): 171–83; Hansen, C. H., and Hansen, R. D., "Finding the Face in the Crowd: An Anger Superiority Effect," *Journal of Personality and Social Psychology* 54, no. 6 (1988): 917–24; Baumeister et al., "Bad Is Stronger Than Good."

16. Kahneman, D., and Tversky, A., "Choices, Values and Frames," *American Psychologist* 39, no. 4 (1984): 341–50; Kahneman, D., Knetch, J. L., and Thaler, R. H., "Anomalies: The Endowment Effect, Loss Aversion, and Status Quo Bias," *Journal of Economic Perspectives* 5, no. 1 (1991): 193–206; Thaler, R. H., *The Winner's Curse* (Princeton, NJ: Princeton University Press, 1992).

17. Baumeister et al., "Bad Is Stronger Than Good," pp. 323–24; Rozin, P., and Royzman, E. B., "Negativity Bias, Negativity Dominance, and Contagion," *Personality and Social Psychology Review* 5, no. 4 (2001): 296–320.

18. Johnson, D. D. P., *Overconfidence and War: The Havoc and Glory of Positive Illusions* (Cambridge, MA: Harvard University Press, 2004); Sharot, T., *The Optimism Bias: A Tour of The Irrationally Positive Brain* (New York: Pantheon, 2011); Taylor, S. E., *Positive*

Illusions: Creative Self-Deception and the Healthy Mind (New York: Basic Books, 1989); Brown, D. E., *Human Universals* (New York: McGraw-Hill, 1991).

19. Sagarin, R., *Learning From the Octopus: How Secrets from Nature Can Help Us Fight Terrorist Attacks, Natural Disasters, and Disease* (New York: Basic Books, 2012).

20. Vermeij, G. J., *Nature: An Economic History* (Princeton, NJ: Princeton University Press, 2004).

21. Haselton, M. G., and Nettle, D., "The Paranoid Optimist: An Integrative Evolutionary Model of Cognitive Biases," *Personality and Social Psychology Review* 10, no. 1 (2006): 47–66; Nesse, R. M., "Natural Selection and the Regulation of Defenses: A Signal Detection Analysis of the Smoke Detector Problem," *Evolution and Human Behavior* 26 (2005): 88–105; Johnson, D. D. P. et al., "The Evolution of Error: Error Management, Cognitive Constraints, and Adaptive Decision-Making Biases," *Trends in Ecology & Evolution* 28, no. 8 (2013): 474–81.

22. McDermott, R., "The Feeling of Rationality: The Meaning of Neuroscientific Advances for Political Science," *Perspectives on Politics* 2, no. 4 (2004): 691–706, p. 695; Sitkin, S. B., "Learning Through Failure," *Research in Organizational Behavior* 14 (1992): 231–66; Johnson, D. D. P., and Madin, E. M. P., "Paradigm Shifts in Security Strategy: Why Does it Take Disasters to Trigger Change?," in *Natural Security: A Darwinian Approach to a Dangerous World*, ed. Sagarin, R. D., and Taylor, T. (Berkeley: University of California Press, 2008).

23. Wilson, T. D., *Strangers to Ourselves: Discovering the Adaptive Unconscious* (Cambridge, MA: Belknap Press, 2004), pp. 63–64.

24. Kahneman, D., *Thinking, Fast and Slow* (London: Allen Lane, 2011).

25. Darwinian fitness means reproductive success, or the number of surviving offspring one contributes to the next generation.

26. Clutton-Brock, T. H., and Parker, G. A., "Punishment in Animal Societies," *Nature* 373, no. 6511 (1995): 209–16, p. 214.

27. Ibid., p. 215; de Waal, F. B. M., *Chimpanzee Politics: Power and Sex among Apes* (Baltimore, MD: Johns Hopkins University Press, 1998); Flack, J. C. et al., "Policing Stabilizes Construction of Social Niches in Primates," *Nature* 439, no. 7075 (2006): 426–29.

28. Clutton-Brock and Parker, "Punishment in Animal Societies," p. 209.
29. Sigmund, "Punish or Perish?"; Baumard, N., "Has Punishment Played a Role in the Evolution of Cooperation? A Critical Review," *Mind and Society* 9, no. 2 (2010): 171–92.
30. Henrich, J., and Boyd, R., "Why People Punish Defectors: Weak Conformist Transmission Can Stabilize Costly Enforcement of Norms in Cooperative Dilemmas," *Journal of Theoretical Biology* 208, no. 1 (2001): 79–89, p. 80. See also Cosmides, L., and Tooby, J., "Cognitive Adaptations for Social Exchange," in *The Adapted Mind: Evolutionary Psychology and the Generation of Culture*, ed. Barkow, J. H., Cosmides, L., and Tooby, J. (New York: Oxford University Press, 1992).

CHAPTER 3

1. Plato, *The Phaedo*.
2. See www.dec.org.uk/haiti-earthquake-facts-and-figures and www.un.org/en/peacekeeping/missions/minustah.
3. Wood, James, "Between God and a Hard Place," *New York Times* Op-Ed, January 23, 2010; Bhatia, Pooja, "Haiti's Angry God," *New York Times* Op-Ed, January 13, 2010.
4. Burleigh, Marc, "Voodoo Practitioners Shrug Off Blame for Haitian Quake," *The Telegraph*, March 8, 2010; Than, Ker, "Haiti Earthquake & Voodoo: Myths, Ritual, and Robertson," *National Geographic News*, January 25, 2010.
5. Harrison, E., Bromiley, G., and Henry, C., eds., *Wycliffe Dictionary of Theology* (Peabody, MA: Hendrickson, 1960), pp. 196, 430; Wright, R., *The Evolution of God* (New York: Little, Brown, 2009), chapter 11.
6. Bernstein, A. E., *The Formation of Hell: Death and Retribution in the Ancient and Early Christian Worlds* (Ithaca, NY: Cornell University Press, 1993), p. x.
7. Neusner, J., *Judaism: The Basics* (New York: Routledge, 2006), p. 138; Konner, M., *The Evolution of Childhood: Relationships, Emotion, Mind* (Cambridge, MA: Belknap Press, 2011), p. 667. It is debated among rabbis whether the blessing means the father could

formerly be punished for the son's sins or, oppositely, whether the son could formerly be punished for the father's sins. Either way, the boy now faces God's punishment for any future sins of his own. Ibid., p. 683.

8. Abdel Haleem (Translator), M. A. S., *The Qur'an* (New York: Oxford University Press, 2004), (3:104); Ruthven, M., *Islam: A Very Short Introduction* (Oxford: Oxford University Press, 1997), pp. 73, 85; Gabriel, R. A., *Muhammad: Islam's First Great General* (Norman: University of Oklahoma Press 2007); Aslan, R., *No God But God: The Origins, Evolution and Future of Islam* (London: Arrow, 2011).

9. Young, M. J. et al., "Deity and Destiny: Patterns of Fatalistic Thinking in Christian and Hindu Cultures," *Journal of Cross-Cultural Psychology* 42, no. 6 (2011): 1030–53.

10. Gyatso, G. K., *Buddhism in the Tibetan Tradition*, trans. Phunrabpa, T. P. (London: Arkana, 1984), pp. 36–37.

11. Ibid., pp. 40, 34.

12. Kirsch, J., *God Against The Gods: The History of the War Between Monotheism and Polytheism* (New York: Penguin, 2004); Wright, *The Evolution of God.*

13. Warrior, V. M., *Roman Religion* (Cambridge: Cambridge University Press, 2006), pp. 49–50.

14. That is, during the Pleistocence period from around 2 million years ago up until 10,000 years ago—which roughly corresponds to the end of the last ice age and the beginning of agriculture.

15. Dudley, M. K., *A Hawaiian Nation I: Man, Gods, and Nature* (Kapolei, HI: Na Kane O Ka Malo Press, 2003), p. 33. An *akua* thus "knows and wills, has capabilities for desires and emotions, and also has capabilities for the extrasensory powers of telepathy, clairvoyance, precognition, and psychokinesis."

16. Ibid., p. 3.

17. Káne, H. K., *Ancient Hawai'i* (Captain Cook, HI: Kawainui Press, 1997), p. 29.

18. Dudley, *A Hawaiian Nation I: Man, Gods, and Nature*, p. 97.

19. Hood, B. M., *SuperSense: Why We Believe in the Unbelievable* (London: HarperOne, 2009), p. 2; Dudley, *A Hawaiian Nation I: Man, Gods, and Nature*, pp. 103–104.

20. Dudley, *A Hawaiian Nation I: Man, Gods, and Nature*, pp. 51, 82.

21. Schneider, D. M., "Political Organization, Supernatural Sanctions and the Punishment for Incest on Yap," *American Anthropologist* 59, no. 5 (1957): 791–800, p. 797.

22. Hadnes, M., and Schumacher, H., "The Gods Are Watching: An Experimental Study of Religion and Traditional Belief in Burkina Faso," *Journal for the Scientific Study of Religion* 51, no. 4 (2012): 689–704, p. 689 ("mostly involve"), p. 92 ("People believe"), p. 91 ("land of honest men"); Mbiti, J. S., *African Religions and Philosophy* (Oxford: Heinemann, 1990), p. 200.

23. Hultkrantz, A., *The Religions of the American Indians* (Berkeley: University of California Press, 1967), p. 18.

24. Swanson, G. E., *The Birth of the Gods* (Ann Arbor: University of Michigan Press, 1960). See also Earhart, H. B., ed. *Religious Traditions of the World* (New York: Harper Collins, 1993).

25. Murdock, G. P., *Theories of Illness: A World Survey* (Pittsburg, PA: University of Pittsburgh Press, 1980).

26. Brown, D. E., *Human Universals* (New York: McGraw-Hill, 1991); Whitehouse, H., "Cognitive Evolution and Religion; Cognition and Religious Evolution," in *The Evolution of Religion: Studies, Theories, and Critiques*, ed. Bulbulia, J., et al. (Santa Margarita, CA: Collins Foundation Press, 2008). The rest of Whitehouse's twelve characteristics were: ritual exegesis, the sacred, signs and portents, deference, creationism, spirit possession, rituals, and revelation.

27. Johnson, D. D. P., "God's Punishment and Public Goods: A Test of the Supernatural Punishment Hypothesis in 186 World Cultures," *Human Nature* 16, no. 4 (2005): 410–46.

28. Ostrom, E., *Governing the Commons: The Evolution of Institutions for Collective Action* (Cambridge: Cambridge University Press, 1990); Lansing, J. S., *Perfect Order: Recognizing Complexity in Bali* (Princeton, NJ: Princeton University Press, 2006).

29. Hartberg, Y. et al., "Supernatural Monitoring and Sanctioning in Community-Based Resource Management," *Religion, Brain & Behavior* (in press; published online ahead of print version by Taylor & Francis, October 13, 2014, DOI:10.1080/2153599X.2014.959547).

30. Ibid., p. 13 (of online manuscript).

31. Pew Research Center's 2007 "U.S. Religious Landscape Survey," http://religions.pewforum.org. The Baylor Religion Survey in 2006 found similar results, for example that 63 percent of Americans "not affiliated with a religious tradition believe in God or some higher power." www.baylorisr.org.

32. Wave II of the survey in 2007 found similar results: 49 percent did not think it described God very well or at all, while 38 percent thought it described God somewhat well or very well (the remainder were undecided).

33. These data were calculated from the World Values Survey aggregated data for 1999/2000, available from www.worldvaluessurvey.org.

34. Atkinson, Q. D., and Bourrat, P., "Beliefs About God, the Afterlife and Morality Support the Role of Supernatural Policing in Human Cooperation," *Evolution and Human Behavior* 32, no. 1 (2011): 41–49; Shariff, A. F., and Rhemtulla, M., "Divergent Effects of Beliefs in Heaven and Hell on National Crime Rates," *Public Library of Science ONE (PLoS ONE)* 7, no. 6 (2012): e39048.

35. Shariff, A. F., and Norenzayan, A., "God Is Watching You: Supernatural Agent Concepts Increase Prosocial Behavior in an Anonymous Economic Game," *Psychological Science* 18, no. 9 (2007): 803–809. For background on economic games, see Kagel, J. H., and Roth, A. E., eds., *The Handbook of Experimental Economics* (Princeton, NJ: Princeton University Press, 1995).

36. Shariff, A. F., and Norenzayan, A., "Mean Gods Make Good People: Different Views of God Predict Cheating Behavior," *International Journal for the Psychology of Religion* 21, no. 2 (2011): 85–96; Purzycki, B. et al., "What Does God Know? Supernatural Agents' Access to Socially Strategic and Nonstrategic Information," *Cognitive Science* 36, no. 5 (2012): 846–69.

37. McKay, R. T. et al., "Wrath of God: Religious Primes and Punishment," *Proceedings of the Royal Society B: Biological Sciences* 278, no. 1713 (2010): 1858–63.

38. Henrich, J., Heine, S., and Norenzayan, A., "The Weirdest People in the World?," *Behavioral and Brain Sciences* 33, no. 2–3 (2010): 61–83; Hadnes and Schumacher, "The Gods Are Watching," p. 702.

CHAPTER 4

1. Citations from the *Holy Bible*, New Revised Standard Edition (Cambridge: Cambridge University Press, 1989). Exodus 20:18; Leviticus 26.

2. Johnson, D. D. P., "Why God Is the Best Punisher," *Religion, Brain & Behavior (RBB)* 1, no. 1 (2011): 77–84.

3. Sosis, R., and Bressler, E. R., "Cooperation and Commune Longevity: A Test of the Costly Signaling Theory of Religion," *Cross-Cultural Research* 37, no. 2 (2003): 211–39; Sosis, R., and Ruffle, B., "Religious Ritual and Cooperation: Testing for a Relationship on Israeli Religious and Secular Kibbutzim," *Current Anthropology* 44, no. 5 (2003): 713–22; Norenzayan, A., *Big Gods: How Religion Transformed Cooperation and Conflict* (Princeton, NJ: Princeton University Press, 2013); Wilson, D. S., *Darwin's Cathedral: Evolution, Religion, and the Nature of Society* (Chicago: University of Chicago Press, 2002).

4. Thomas, W. I., and Thomas, D. S., *The Child in America: Behaviour Problems and Programs* (New York: Knopf, 1928).

5. Schneider, D. M., "Political Organization, Supernatural Sanctions and the Punishment for Incest on Yap," *American Anthropologist* 59, no. 5 (1957): 791–800, p. 798.

6. Pettitt, P., *The Palaeolithic Origins of Human Burial* (New York: Routledge, 2011), pp. 59–62, 62; Ronen, A., "The Oldest Burials and Their Significance," in *African Genesis: Perspectives on Hominin Evolution*, ed. Reynolds, S. C., and Gallagher, A. (2012). We have to be careful interpreting these earliest burial goods. Pettitt doesn't mention religion anywhere in his entire book, only noting that "grave goods may or may not be related to metaphysical notions of an afterlife or bodily extension; they probably speak more of self-expression and concepts of ownership" (p. 129). Since we know afterlife beliefs eventually did emerge, however, they had to begin *at some point*, and they may thus logically be as old as some of the oldest known examples of burial goods.

7. Leakey, R., and Lewin, R., *Origins Reconsidered: In Search of What Makes Us Human* (London: Anchor, 1993).

8. Budge, E. A. W. T., and Romer, J. I., *The Egyptian Book of the Dead*, Penguin Classics (London: Penguin, [1899] 2008); Deborah

Mackenzie, "Pharaoh's Pots Give Up Their Secrets," *New Scientist*, March 17, 2007.

9. Baumard, N., and Boyer, P., "Explaining Moral Religions," *Trends in Cognitive Sciences* 17, no. 6 (2013): 272–80, p. 274.

10. Brown, D. E., *Human Universals* (New York: McGraw-Hill, 1991); Whitehouse, H., "Cognitive Evolution and Religion; Cognition and Religious Evolution," in *The Evolution of Religion: Studies, Theories, and Critiques*, ed. Bulbulia, J., et al. (Santa Margarita, CA: Collins Foundation Press, 2008); Hultkrantz, A., *The Religions of the American Indians* (Berkeley: University of California Press, 1967), pp. 130, 37.

11. Hultkrantz, *The Religions of the American Indians*, pp. 135–36.

12. Ibid., pp. 131–33; Earhart, H. B., ed. *Religious Traditions of the World* (New York: Harper Collins, 1993).

13. Hultkrantz, *The Religions of the American Indians*, pp. 132–33.

14. Ibid., p. 133.

15. Ibid., p. 133 ("different passageways"), p. 27 ("recognize a heaven"), p. 27 ("all developed from"), p. 27, fn 1 ("This threefold pattern").

16. Peoples, H. C., and Marlowe, F. W., "Subsistence and the Evolution of Religion," *Human Nature* 23, no. 3 (2012): 253–69.

17. Hultkrantz, *The Religions of the American Indians*, pp. 134–35.

18. Harris, S., *The End of Faith: Religion, Terror, and the Future of Reason*, 1st ed. (New York: Norton, 2004), p. 89.

19. Brown, *Human Universals*; Whitehouse, "Cognitive Evolution and Religion"; Swanson, G. E., *The Birth of the Gods* (Ann Arbor: University of Michigan Press, 1960); Boehm, C., "A Biocultural Evolutionary Exploration of Supernatural Sanctioning," in *The Evolution of Religion: Studies, Theories, and Critiques*, ed. Bulbulia, J., et al. (Santa Margarita, CA: Collins Foundation Press, 2008); Hartberg et al., Y., "Supernatural Monitoring and Sanctioning in Community-Based Resource Management," *Religion, Brain & Behavior* (in press; published online ahead of print version by Taylor & Francis, October 13, 2014, DOI:10.10 80/2153599X.2014.959547); Mbiti, J. S., *African Religions and Philosophy* (Oxford: Heinemann, 1990); Hultkrantz, *The Religions of the American Indians*, p. 129.

20. Hultkrantz, *The Religions of the American Indians*, p. 88.

21. Ibid. My italics.

22. Hadnes, M., and Schumacher, H., "The Gods Are Watching: An Experimental Study of Religion and Traditional Belief in Burkina Faso," *Journal for the Scientific Study of Religion* 51, no. 4 (2012): 689–704; Mbiti, *African Religions and Philosophy*; Bering, J. M., and Johnson, D. D. P., " 'Oh Lord, You Hear My Thoughts from Afar': Recursiveness in the Cognitive Evolution of Supernatural Agency," *Journal of Cognition and Culture* 5, no. 1/2 (2005): 118–42; Hultkrantz, *The Religions of the American Indians*, p. 138 ("the dead occasion"); Rossano, M. J., "Supernaturalizing Social Life: Religion and the Evolution of Human Cooperation," *Human Nature* 18, no. 3 (2007): 272–94, p. 280 ("ever watchful, active players").

23. Rossano, "Supernaturalizing Social Life," p. 279.

24. Ibid., p. 284.

25. Hultkrantz, *The Religions of the American Indians*, p. 12.

26. Ibid., p. 72 ("each individual possess"); ibid., p. 124 ("pledge that the"); ibid., p. 127 ("cult room"); Hadnes and Schumacher, "The Gods Are Watching," p. 689 ("consult with witch doctors"); Rossano, "Supernaturalizing Social Life," p. 280 ("ancestors may simply"); Dudley, M. K., *A Hawaiian Nation I: Man, Gods, and Nature* (Kapolei, HI: Na Kane O Ka Malo Press, 2003).

27. Pals, D. L., *Eight Theories of Religion* (New York: Oxford University Press, 2006); Cronk, L., "Evolutionary Theories of Morality and the Manipulative Use of Signals," *Zygon* 29, no. 1 (1994): 81–101; Dawkins, R., *The God Delusion* (New York: Houghton Mifflin, 2006); Wilson, *Darwin's Cathedral: Evolution, Religion, and the Nature of Society*; Sosis, R., "Why Aren't We All Hutterites? Costly Signaling Theory and Religious Behavior," *Human Nature* 14, no. 2 (2003): 91–127; Whitehouse, H., *Modes of Religiosity: A Cognitive Theory of Religious Transmission* (Walnut Creek, CA: Altamira Press, 2004); Whitehouse, H., *Arguments and Icons: Divergent Modes of Religiosity* (Oxford: Oxford University Press, 2000).

28. Shariff, A. F., and Rhemtulla, M., "Divergent Effects of Beliefs in Heaven and Hell on National Crime Rates," *Public Library of Science ONE (PLoS ONE)* 7, no. 6 (2012): e39048, p. 1.

29. Sosis, R., Kress, H., and Boster, J., "Scars for War: Evaluating Alternative Signaling Explanations for Cross-Cultural Variance in Ritual Costs," *Evolution and Human Behavior* 28, no. 4 (2007): 234–47; Snarey, J., "The Natural Environment's Impact Upon Religious Ethics: A Cross-Cultural Study," *Journal for the Scientific Study of Religion* 35, no. 2(1996): 85–96; Wilson, D. S., "Testing Major Evolutionary Hypotheses about Religion with a Random Sample," *Human Nature* 16, no. 4 (2005): 419–46.

30. Beal, T. K., *Religion and Its Monsters* (New York: Routledge, 2002).

31. Some commonalities do remain surprising in their specificity. For example, the legend of a great flood is widespread in indigenous cultures as well, stretching across the Americas (Hultkrantz, *The Religions of the American Indians*, p. 30). Even the details are similar. Some time long ago, "all men except for a couple of individuals were drowned in a great deluge which covered all the land. The two who escaped took refuge on a raft or an oak log (common in North American versions) or on top of a mountain or a palm tree (frequent in the South American legends)"; ibid., p. 30. As in the Biblical version, these floods were retribution for man's misdeeds, "for example, by the wrath of a god or the transgression of a taboo"; ibid., p. 31.

32. Baumard and Boyer, "Explaining Moral Religions"; Norenzayan, *Big Gods*; Brown, *Human Universals*; Rossano, "Supernaturalizing Social Life," p. 280; Malinowski, B., *The Foundations of Faith and Morals: An Anthropological Analysis of Primitive Beliefs and Conduct with Special Reference to the Fundamental Problem of Religion and Ethics* (London: Oxford University Press, 1935), p. viii.

33. Welch, D. A., *Justice and the Genesis of War* (Cambridge: Cambridge University Press, 1993), p. 195; Hauser, M. D., *Moral Minds: How Nature Designed our Universal Sense of Right and Wrong*, 1st ed. (New York: Ecco, 2006); Haidt, J., "The New Synthesis in Moral Psychology," *Science* 316, no. 5827 (2007): 998–1002; Bloom, P., *Just Babies: The Origins of Good and Evil* (New York: Crown, 2013).

34. Aku Visala, personal communication (2013); Voltaire, "A l'Auteur du Livre des Trois Imposteurs," *Épîtres* No. 96 (1769), p. 229.

CHAPTER 5

1. Shakespeare, William, *Henry VI*, Part III, Act V, scene 6.
2. *The Times* (London), "How Did You Help England Win the Ashes?," September 16, 2005.
3. Hood, *SuperSense: Why We Believe in the Unbelievable*, p. 8. Hood, Bruce, "Take the Ladder Test," *The Guardian Weekend Magazine*, May 9, 2009, p. 27. Gross, N., and Simmons, S., "The Religiosity of American College and University Professors," *Sociology of Religion* 70, no. 2 (2010): 101–29; Wilson, S., "The Naturalness of Weird Beliefs," *The Psychologist* 23, no. 7 (2010): 564–67, p. 566.
4. Baker, W. J., *Playing with God: Religion and Modern Sport* (Cambridge, MA: Harvard University Press, 2007).
5. Solomon, Alan, "Cubs Devouring Wendell's Act," *Chicago Tribune*, March 11, 1993.
6. Burger, J. M., and Lynn, A. L., "Superstitious Behavior Among American and Japanese Professional Baseball Players," *Basic and Applied Social Psychology* 27, no. 1 (2005): 71–76, p. 74. For a famous earlier study of superstition in baseball, see Gmelch, G., "Superstition and Ritual in American Baseball," *Elysian Fields Quarterly* 11, no. 3 (1992): 25–36; Gmelch, G., "Baseball Magic," *Society* 8, no. 8 (1971): 39–41.
7. Damisch, L., Stoberock, B., and Mussweiler, T., "Keep Your Fingers Crossed! How Superstition Improves Performance," *Psychological Science* 21, no. 7 (2010): 1014–20; Moran, A., *Sport and Exercise Psychology: A Critical Introduction* (New York: Routledge, 2004), p. 117. David White "This is Wimbledon? You Cannot Be Serious," *The Sunday Times Magazine* (London), May 30, 2010, p. 12–17.
8. Hood, *SuperSense*, p. 29.
9. Moran, *Sport and Exercise Psychology*, p. 117.
10. Ibid., p. 118. Original italics.
11. Gallup Poll 2005, www.gallup.com/poll/19558/Paranormal-Beliefs-Come-SuperNaturally-Some.aspx. Survey by Richard Wiseman of 2,068 people for 2003 national science week. Cited in Hood, "Take the Ladder Test," p. 27.
12. Hood, *SuperSense*, p. 50.
13. Ibid., p. 117.

14. Pronin, E. et al., "Everyday Magical Powers: The Role of Apparent Mental Causation in the Overestimation of Personal Influence," *Journal of Personality and Social Psychology* 91 (2006): 218–31; Shermer, M., *Why People Believe Weird Things: Pseudoscience, Superstition, and Other Confusions of Our Time* (Holt, 2002); Barrett, J. L., *Why Would Anyone Believe in God?* (Altamira Press, 2004); Wiseman, R., *Paranormality: Why We See What Isnt' There* (Spin Solutions, 2010).

15. Baylor University Religion Survey 2006, www.baylor.edu/mediacommunications/news.php?action=story&story=41678. Kohut, A. et al., *The Diminishing Divide: Religion's Changing Role in American Politics* (Washington, DC: Brookings Institution Press, 2000), p. 29 ("34 percent of atheists"). Pew Forum U.S. Religious Landscape Survey, http://religions.pewforum.org conducted May 8 to August 13, 2007, among more than 35,000 Americans age 18 and older. www.pewforum.org/Not-All-Nonbelievers-Call-Themselves-Atheists.aspx.

16. Johnson, D. D. P., and Reeve, Z., "The Virtues of Intolerance: Is Religion an Adaptation for War?," in *Religion, Intolerance and Conflict: A Scientific and Conceptual Investigation*, ed. Clarke, S., Powell, R., and Savulescu, J. (Oxford: Oxford University Press, 2013); Johnson, D. D. P., "Gods of War: The Adaptive Logic of Religious Conflict," in *The Evolution of Religion: Studies, Theories, and Critiques*, ed. Bulbulia, J., et al. (Santa Margarita, CA: Collins Foundation Press, 2008); MacNeill, A., "The Capacity for Religious Experience is an Evolutionary Adaptation to Warfare," *Evolution and Cognition* 10, no. 1 (2004): 43–60.

17. Newhouse, Alana, "At 36,000 Feet, Closer to God," March 14, 2008, *New York Times*, p. D4. Robin Marantz Henig, "Darwin's God," *New York Times Magazine*, March 4, 2007.

18. Sosis, R., and Handwerker, W. P., "Psalms and Coping with Uncertainty: Israeli Women's Responses to the 2006 Lebanon War," *American Anthropologist* 113, no. 1 (2011): 40–55; Hood, *SuperSense*, pp. 31–33; Seligman, M. E. P., "Learned Helplessness," *Annual Review of Medicine* 23, no. 1 (1972): 407–12.

19. Sosis, R., and Handwerker, W. P., "Psalms and Coping with Uncertainty: Israeli Women's Responses to the 2006 Lebanon War,"

p. 42 ("are believed to"), p. 40 ("Jews, don't just"), p. 44 ("traveling, depression, healing").

20. Ibid., pp. 42–43.

21. Keinan, G., "The Effects of Stress and Desire for Control on Superstitious Behavior," 28, no. 1 (2002): 102–10.

22. Malinowski, B., *Argonauts of the Western Pacific* (Prospect Heights, IL: Waveland Press, 1961); Sosis and Handwerker, "Psalms and Coping with Uncertainty"; Whitson, J. A., and Galinsky, A. D., "Lacking Control Increases Illusory Pattern Perception," *Science* 322, no. 5898 (2008): 115–17.

23. Pisacreta, R., "Superstitious Behavior and Response Stereotypy Prevent the Emergence of Efficient Rule-Governed Behavior in Humans," *Psychological Record* 48, no. 2 (1998): 251–74; Foster, K. R., and Kokko, H., "The Evolution of Superstitious and Superstitious-Like Behaviour," *Proceedings of the Royal Society B: Biological Sciences* 276, no. 1654 (2009): 31–37, p. 31.

24. Foster and Kokko, "The Evolution of Superstitious and Superstitious-Like Behaviour," pp. 31–32. On obsessive behavior: Boyer, P., and Liernard, P., "Ritual Behavior in Obsessive and Normal Individuals," *Current Directions in Psychological Science* 17, no. 4 (2008): 291–94. On erring on the side of caution: Beck, J., and Forstmeier, W., "Superstition and Belief as Inevitable By-Products of an Adaptive Learning Strategy," *Human Nature* 18, no. 1 (2007): 35–46; Johnson, D. D. P. et al., "The Evolution of Error: Error Management, Cognitive Constraints, and Adaptive Decision-Making Biases," *Trends in Ecology & Evolution* 28, no. 8 (2013): 474–81; Haselton, M. G., and Nettle, D., "The Paranoid Optimist: An Integrative Evolutionary Model of Cognitive Biases," *Personality and Social Psychology Review* 10, no. 1 (2006): 47–66.

25. Hume, D., *The Natural History of Religion* (London: A. and H. Bradlaugh Bonner, 1889/1757); full text available at http://oll.libertyfund.org/titles/340.

26. Ibid., p. 8.

27. Hood, "Take the Ladder Test," p. 30.

28. Hood, *SuperSense*, p. 7.

29. Converse, B. A., Risen, J. L., and Carter, T. J., "Investing in Karma: When Wanting Promotes Helping," *Psychological Science* 23, no. 8 (2012): 923–30, p. 923.

30. Hood, *SuperSense*, p. 10; Gilovich, T., Griffin, D., and Kahneman, D., eds., *Heuristics and Biases: The Psychology of Intuitive Judgment* (Cambridge: Cambridge University Press, 2002); Fiske, S. T., and Taylor, S. E., *Social Cognition: From Brains to Culture* (New York: McGraw-Hill, 2007); Kahneman, D., *Thinking, Fast and Slow* (London: Allen Lane, 2011).

31. Atran, S., *In Gods We Trust: The Evolutionary Landscape of Religion* (Oxford: Oxford University Press, 2004); Barrett, J., *Born Believers: The Science of Children's Religious Beliefs* (Atria Books, 2012); Boyer, P., *Religion Explained: The Evolutionary Origins of Religious Thought* (New York: Basic Books, 2001).

32. Hood, *SuperSense*, p. 62.

33. Foster and Kokko, "The Evolution of Superstitious and Superstitious-Like Behaviour"; Shermer, M., "Wheatgrass Juice and Folk Medicine," *Scientific American* 299, no. 2 (2008): 42.

34. Bering, J. M., "The Existential Theory of Mind," *Review of General Psychology* 6 (2002): 3–24.

35. Choi, I., Nisbett, R. E., and Norenzayan, A., "Causal Attribution Across Cultures: Variation and Universality," *Psychological Bulletin* 125, no. 1 (1999): 47–63; Nisbett, R. E., *The Geography of Thought: How Asians and Westerners Think Differently—and Why* (Yarmouth, ME: Nicholas Brealey Publishing, 2003).

36. Wegner, D. M., *The Illusion of Conscious Will* (Cambridge, MA: MIT Press, 2002).

37. Bering, J. M., "The Folk Psychology of Souls," *Behavioural and Brain Sciences* 29, no. 5 (2006): 453–62; Bloom, P., *Descartes' Baby: How the Science of Child Development Explains What Makes Us Human* (New York: Basic Books, 2004).

38. Bering, J. M., and Bjorklund, D. F., "The Natural Emergence of Reasoning About the Afterlife as a Developmental Regularity," *Developmental Psychology* 40, no. 2 (2004): 217–33. Though see work by Mitch Hodge that questions some of these findings: Hodge, K. M., "Context Sensitivity and the Folk Psychology of Souls: Why

Bering et. al. Got the Findings They Did," in *Is Religion Natural?*, ed. Evers, D., et al. (New York: T&T Clark International, 2012).

39. Abell, F., Happe, F., and Frith, U., "Do Triangles Play Tricks? Attributions of Mental States to Animated Shapes in Normal and Abnormal Development," *Cognitive Development* 15, no. 1 (2000): 1–16. See also Guthrie, S. E., *Faces in the Clouds: A New Theory of Religion* (New York: Oxford University Press, 1993); Barrett, *Why Would Anyone Believe in God?*

40. Bering, J. M., McLeod, K. A., and Shackelford, T. K., "Reasoning About Dead Agents Reveals Possible Adaptive Trends," *Human Nature* 16, no. 4 (2005): 360–81.

41. Burnham, T., and Hare, B., "Engineering Cooperation: Does Involuntary Neural Activation Increase Public Goods Contributions?," *Human Nature* 18, no. 2 (2007): 88–108; Bateson, M., Nettle, D., and Roberts, G., "Cues of Being Watched Enhance Cooperation in a Real-World Setting," *Biology Letters* 2, no. 3 (2006): 412–14; Haley, K., and Fessler, D., "Nobody's Watching? Subtle Cues Affect Generosity in an Anonymous Economic Game," *Evolution and Human Behavior* 26, no. 3 (2005): 245–56; Rigdon, M. et al., "Minimal Social Cues in the Dictator Game," *Journal of Economic Psychology* 30, no. 3 (2009): 358–67.

42. Morewedge, C. K., "Negativity Bias in Attribution of External Agency," *Journal of Experimental Psychology* 138, no. 4 (2009): 535–45; Baumeister, R. F. et al., "Bad Is Stronger Than Good," *Review of General Psychology* 5, no. 4 (2001): 323–70.

43. Andre, C., and Velasquez, M., "Just World Theory," *Issues in Ethics* 3, no. 2 (1990), http://www.scu.edu/ethics/publications/iie/v3n2/justworld.html.

44. Lerner, M. J., and Miller, D. T., "Just World Research and the Attribution Process: Looking Back and Ahead," *Psychological Bulletin* 85, no. 5 (1978): 1030–1051; Lerner, M. J., *The Belief in a Just World: A Fundamental Delusion* (New York: Plenum Press, 1980); Montada, L., and Lerner, M. J., eds., *Responses to Victimization and Belief in a Just World* (New York: Plenum Press, 2010); Hafer, C. L., and Bègue, L., "Experimental Research on Just-World Theory: Problems, Developments, and Future Challenges," *Psychological*

Bulletin 131, no. 1 (2005): 128–67; Piaget, J., *The Moral Judgment of the Child* (New York: Free Press, 1997).

45. Callan, M. J., Ellard, J. H., and Nicol, J. E., "The Belief in a Just World and Immanent Justice Reasoning in Adults," *Personality and Social Psychology Bulletin* 32, no. 12 (2006): 1646–58, p. 1646.

46. Anderson, J. E., Kay, A. C., and Fitzsimons, G. M., "In Search of the Silver Lining: The Justice Motive Fosters Perceptions of Benefits in the Later Lives of Tragedy Victims," *Psychological Science* 21, no. 11 (2010): 1599–604, p. 1599.

47. Callan, M. J. et al., "Gambling as a Search for Justice: Examining the Role of Personal Relative Deprivation in Gambling Urges and Gambling Behavior," *Personality and Social Psychology Bulletin* 34, no. 11 (2008): 1514–29.

48. Gaucher, D. et al., "Compensatory Rationalizations and the Resolution of Everyday Undeserved Outcomes," *Personality and Social Psychology Bulletin* 36, no. 1 (2010): 109–18.

49. Kaplan, H., "Belief in a Just World, Religiosity and Victim Blaming," *Archive for the Psychology of Religion / Archiv für Religionspychologie* 34, no. 3 (2012): 397–409.

50. Rice, S. et al., "Damned If You Do and Damned If You Don't: Assigning Blame to Victims Regardless of Their Choice," *Social Science Journal* 49, no. 1 (2012): 5–8, p. 5.

51. Bloom, *Descartes' Baby*; Bering, "The Folk Psychology of Souls"; Bering, J. M., and Parker, B. D., "Children's Attributions of Intentions to an Invisible Agent," *Developmental Psychology* 42, no. 2 (2006): 253–62; Alcorta, C. S., and Sosis, R., "Ritual, Emotion, and Sacred Symbols: The Evolution of Religion as an Adaptive Complex," *Human Nature* 16, no. 4 (2005): 323–59; Barrett, *Born Believers*; Keleman, D., "Are Children 'Intuitive Theists'? Reasoning About Purpose and Design in Nature," *Psychological Science* 15, no. 5 (2004): 295–301; Bering and Bjorklund, "The Natural Emergence of Reasoning About the Afterlife as a Developmental Regularity"; Bulbulia, J., "Meme Infection or Religious Niche Construction? An Adaptationist Alternative to the Cultural Maladaptationist Hypothesis," *Method and Theory in the Study of Religion* 20, no. 1 (2008): 67–107, p. 96. Some are skeptical of the teleological reasoning claim, since it might simply be that children, and adults in

prescientific societies, simply endorse a simpler view rather than a more complex one.

52. Dawkins, R., *The God Delusion* (New York: Houghton Mifflin, 2006); Glaeser, E. L., and Sacerdote, B. I., "Education and Religion," *Journal of Human Capital* 2, no. 2 (2008): 188–215; Lee, L., and Bullivant, S., "Where Do Atheists Come From?," *New Scientist*, no. 6 March (2010): 26–27; Beit-Hallahmi, B., "Atheists: A Psychological Profile," in *The Cambridge Companion to Atheism*, ed. Martin, M. (Cambridge: Cambridge University Press, 2007), p. 308.

53. Bainbridge, W. S., "Atheism," *Interdisciplinary Journal of Research on Religion* 1, no. 1 (2005): Article 2, p.4; Caldwell-Harris, C. L., "Understanding Atheism/Non-Belief as an Expected Individual-Differences Variable," *Religion, Brain & Behavior* 2, no. 1 (2012): 4–23, p. 4; ibid.; Beit-Hallahmi, "Atheists"; Galen, L., "Profiles of the Godless," *Free Inquiry* 29, no. 5 (2009): 41–45.

54. Johnson, D. D. P., "What Are Atheists For? Hypotheses on the Functions of Non-Belief in the Evolution of Religion," *Religion, Brain & Behavior* 2, no. 1 (2012): 48–70; McCauley, R. N., *Why Religion is Natural and Science is Not* (Oxford: Oxford University Press, 2011).

55. There are numerous slightly different versions of this story, one of which can be found in Pais, A., *Inward Bound: Of Matter and Forces in the Physical World* (New York: Oxford University Press, 1988), p. 210.

56. Hood, "Take the Ladder Test," p. 27 ("the number one" and "For believers, examples"). On perceiving and believing: Henriksen, J.-O. *Life, Love and Hope: God and Human Experience* (Grand Rapids, MI: Eerdmans, 2014). On alien abduction: Clancy, S. A., *Abducted: How People Come to Believe They Were Kidnapped by Aliens* (Cambridge, MA: Harvard University Press, 2007). Hood, *SuperSense*, p. 36 ("Even if you"). Fuentes, A., *Race, Monogamy, and Other Lies They Told You: Busting Myths about Human Nature* (Berkely: University of California Press, 2012), p. 27 ("I would not").

57. Hoggard-Creegan, N., "Are Humans Adaptive for the God Niche? An Argument from Mathematics," *Philosophy, Theology and the Sciences* 1, no. 2 (2014): 232–50, p. 234.

58. Hood, "Take the Ladder Test," p. 27.

59. Boyer, *Religion Explained*; Bloom, *Descartes' Baby*; Kirkpatrick, L. A., *Attachment, Evolution, and the Psychology of Religion* (New York: Guilford Press, 2005); Dawkins, *The God Delusion*.

CHAPTER 6

1. Seneca, L. A. *Seneca's Morals*, ed. L'estrange, R. (London: S. Crowder, R. Baldwin, and B. Collins, 1793), p. 26.

2. Stein, S. J., *The Shaker Experience in America: A History of the United Society of Believers* (New Haven, CT: Yale University Press, 1994). Quotes are from Chase, Stacey, "The Last Ones Standing," *Boston Globe*, July 23, 2006.

3. Chase, Stacey, "He Left the Shakers for Love," *Boston Globe Sunday Magazine*, February 28, 2010. All other quotes are from the original 2006 article.

4. Bentley, R., Hahn, M., and Shennan, S., "Random Drift and Culture Change," *Proceedings of the Royal Society B: Biological Sciences* 271, no. 1547 (2004): 1443–50; Wilson, D. S., "Evolution and Religion: The Transformation of the Obvious," in *The Evolution of Religion: Studies, Theories, and Critiques*, ed. Bulbulia, J., et al. (Santa Margarita, CA: Collins Foundation Press, 2008); Diamond, J., *Collapse: How Societies Choose to Fail or Succeed* (New York: Penguin, 2005).

5. Bulbulia, J., "Spreading Order: Religion, Cooperative Niche Construction, and Risky Coordination Problems," *Biology & Philosophy* 27, no. 1 (2012): 1–27; Williams, G. C., *Adaptation and Natural Selection* (Princeton, NJ: Princeton University Press, 1966); Wilson, D. S., and Wilson, E. O., "Rethinking the Theoretical Foundation of Sociobiology," *Quarterly Review of Biology* 82, no. 4 (2007): 327–48; Wilson, D. S., and Sober, E., "Reintroducing Group Selection to the Human Behavioural Sciences," *Behavioral and Brain Sciences* 17, no. 4 (1994): 585–654; Okasha, S., *Evolution and the Levels of Selection* (Oxford: Oxford University Press, 2006); Maynard Smith, J., and Szathmary, E., *The Major Transitions of Life* (New York: W. H. Freeman, 1995).

6. For recent criticism and responses, see Pinker, S., "The False Allure of Group Selection," *Edge* July (2012), http://edge.org/conversation/ the-false-allure-of-group-selection.

7. On apparent altruism in nature: Wilson, E. O., *Sociobiology: The New Synthesis* (Harvard: Belknap Press, 1975); Dawkins, R., *The Selfish Gene* (Oxford: Oxford University Press, 1976); Dugatkin, L. A., *Cooperation in Animals* (Oxford: Oxford University Press, 1997). On new cognitive abilities: Bering, J. M., "The Existential Theory of Mind," *Review of General Psychology* 6, no. 1 (2002): 3–24; Bering, J. M., and Shackelford, T., "The Causal Role of Consciousness: A Conceptual Addendum to Human Evolutionary Psychology," *Review of General Psychology* 8, no. 4 (2004): 227–48. On theory of mind: Povinelli, D. J., and Bering, J. M., "The Mentality of Apes Revisited," *Current Directions in Psychological Science* 11, no. 4 (2002): 115–19. On complex language: Pinker, S., *The Blank Slate: The Modern Denial of Human Nature* (New York: Penguin Putnam, 2002).

8. Dreber, A. et al., "Winners Don't Punish," *Nature* 452, no. 7185 (2008): 348–51, p. 350. On privacy and the Mehinacu example: Locke, J. L., *Eavesdropping: An Intimate History* (Oxford: Oxford University Press, 2010).

9. Barash, David, "The Social Responsibility in Teaching Sociobiology," *Chronicle of Higher Education*, November 17, 2006. Bering and Shackelford, "The Causal Role of Consciousness."

10. Dunbar, R. I. M., *Grooming, Gossip and the Evolution of Language* (London: Faber & Faber, 1996), p. 79.

11. Nikiforakis, N., and Normann, H.-T., "A Comparative Statics Analysis of Punishment in Public-Good Experiments," *Experimental Economics* 11, no. 4 (2008): 358–69; Johnson, D. D. P., and MacKay, N. J., "Fight the Power: Lanchester's Laws of Combat in Human Evolution" *Evolution and Human Behavior* 36, no. 2 (2015): 152–63.

12. Boehm, C., *Hierarchy in the Forest: The Evolution of Egalitarian Behavior* (Cambridge, MA: Harvard University Press, 2001); Boehm, C., *Moral Origins: The Evolution of Virtue, Altruism, and Shame* (New York: Basic Books, 2012); Wrangham, R. W.,

and Peterson, D., *Demonic Males: Apes and the Origins of Human Violence* (London: Bloomsbury, 1996).

13. Baumard, N., "Has Punishment Played a Role in the Evolution of Cooperation? A Critical Review," *Mind and Society* 9, no. 2 (2010): 171–92.

14. Káne, H. K., *Ancient Hawai'i* (Captain Cook, HI: Kawainui Press, 1997), p. 36.

15. Boehm, *Hierarchy in the Forest: The Evolution of Egalitarian Behavior*; Boehm, C. H., *Blood Revenge: The Enactment and Management of Conflict in Montenegro and Other Tribal Societies* (Philadelphia: University of Pennsylvania Press, 1986); Wilson, S., *Feuding, Conflict, and Banditry in Nineteenth-Century Corsica,* (Cambridge: Cambridge University Press, 1988); Wiessner, P., "Norm Enforcement Among the Ju/'hoansi Bushmen: A Case of Strong Reciprocity?," *Human Nature* 16, no. 2 (2005): 115–45.

16. Sigmund, K., "Punish or Perish? Retaliation and Collaboration Among Humans," *Trends in Ecology & Evolution* 22, no. 11 (2007): 593–600, p. 598; Riskey, D. R., and Birnbaum, M. H., "Compensatory Effects in Moral Judgment: Two Rights Don't Make Up for a Wrong," *Journal of Experimental Psychology* 103, no. 1 (1974): 171–73.

17. Dunbar, *Grooming, Gossip and the Evolution of Language*, p. 112.

18. Locke, *Eavesdropping*.

19. Nowak, M., and Sigmund, K., "Evolution of Indirect Reciprocity," *Nature* 437, no. 7063 (2005): 1291–98; Wedekind, C., and Milinski, M., "Cooperation Through Image Scoring in Humans," *Science* 288, no. 5467 (2000): 850–52; van Vugt, M., and Tybur, J. M., "The Evolutionary Foundations of Status and Hierarchy: Dominance, Prestige, Power, and Leadership," in *Handbook of Evolutionary Psychology*, 2nd edn., ed. Buss, D. M. (New York: Wiley, 2015); Henrich, J., and Gil-White, F., "The Evolution of Prestige: Freely Conferred Status as a Mechanism for Enhancing the Benefits of Cultural Transmission," *Evolution and Human Behavior* 22 (2001): 1–32.

20. Piazza, J., and Bering, J. M., "Concerns About Reputation via Gossip Promote Generous Allocations in an Economic Game," *Evolution and Human Behavior* 29, no. 3 (2008): 172–78.

21. Káne, *Ancient Hawai'i*, p. 41.

22. Ghiselin, M. T., "Darwin and Evolutionary Psychology," *Science* 179, no. 4077 (1973): 964–68, p. 967; Dawkins, R., *Unweaving the Rainbow: Science, Delusion and the Appetite for Wonder* (London: Penguin Books, 1998), p. 212.

23. On evolutionarily rooted mechanisms: Barkow, J. H., *Missing the Revolution: Darwinism for Social Scientists* (New York: Oxford University Press, 2006); Buss, D. M., *The Handbook of Evolutionary Psychology* (Hoboken, NJ: Wiley, 2005). On conflicting aspects of the brain: Damasio, A. R., *Descartes Error: Emotion, Reason and the Human Brain* (New York: Avon, 1994); Davidson, R. J., Putnam, K. M., and Larson, C. L., "Dysfunction in the Neural Circuitry of Emotion Regulation: A Possible Prelude to Violence," *Science* 289, no. 5479 (2000): 591–94; Sanfey, A. G. et al., "The Neural Basis of Economic Decision-Making in the Ultimatum Game," *Science* 300, no. 5626 (2003): 1755–58; Lee, M., "Self-Denial and Its Discontents: Toward Clarification of the Intrapersonal Conflict Between 'Selfishness' and 'Altruism,'" in *Evolution, Games and God: The Principle of Cooperation*, ed. Nowak, M. A., and Coakley, S. (Cambridge, MA: Harvard University Press, 2013); Pinker, *The Blank Slate*.

24. Bering and Shackelford, "The Causal Role of Consciousness: A Conceptual Addendum to Human Evolutionary Psychology." On self-control: McCullough, M. E., and Carter, E. C., "Waiting, Tolerating, and Cooperating: Did Religion Evolve to Prop Up Humans' Self-Control Abilities," in *Handbook of Self-Regulation: Research, Theory, and Applications*, ed. Vohs, K. D., and Baumeister, R. F. (New York: Guilford Press, 2011).

25. Goffman, E., *The Presentation of Self in Everyday Life* (New York: Anchor, 1959), pp. 8–9.

26. Iacoboni, M., "Imitation, Empathy, and Mirror Neurons," *Annual Review of Psychology* 60 (2009): 653–70, p. 653.

27. Goldstein, M. A., "The Biological Roots of Heat-of-Passion Crimes and Honor Killings," *Politics and the Life Sciences* 21, no. 2 (2002): 28–37.

28. Barash, David, "The Social Responsibility in Teaching Sociobiology," *Chronicle of Higher Education*, November 17,

2006. The idea of omniscience ("He is always watching") varies across religions, as we saw in Chapter 3, but some form of omniscience and omnipotence in relevant domains is a common theme, especially in popular religion if not in doctrine. Bering, J. M., and Johnson, D. D. P., " 'Oh Lord, You Hear My Thoughts from Afar': Recursiveness in the Cognitive Evolution of Supernatural Agency," *Journal of Cognition and Culture* 5, no. 1/2 (2005): 118–42; Harris, S., *The End of Faith: Religion, Terror, and the Future of Reason*, 1st ed. (New York: Norton, 2004), p. 159. On the development of this supernatural punishment theory: Johnson, D. D. P., "Why God Is the Best Punisher," *Religion, Brain & Behavior (RBB)* 1, no. 1 (2011): 77–84; Johnson, D. D. P., "The Error of God: Error Management Theory, Religion, and the Evolution of Cooperation," in *Games, Groups, and the Global Good*, ed. Levin, S. A. (Berlin: Springer, 2009); Johnson, D. D. P., and Bering, J. M., "Hand of God, Mind of Man: Punishment and Cognition in the Evolution of Cooperation," *Evolutionary Psychology* 4 (2006): 219–33; Johnson, D. D. P., "God's Punishment and Public Goods: A Test of the Supernatural Punishment Hypothesis in 186 World Cultures," *Human Nature* 16, no. 4 (2005): 410–46; Johnson, D. D. P., and Kruger, O., "The Good of Wrath: Supernatural Punishment and the Evolution of Cooperation," *Political Theology* 5, no. 2 (2004): 159–76; Bering and Johnson, " 'Oh Lord, You Hear My Thoughts from Afar.' "

29. Davidson, Putnam, and Larson, "Dysfunction in the Neural Circuitry of Emotion Regulation"; McDermott, R. et al., "Monoamine Oxidase A Gene (MAOA) Predicts Behavioral Aggression Following Provocation," *Proceedings of the National Academy of Sciences* 106, no. 7 (2009): 2118–23.

30. For quote: Hood, B. M., *SuperSense: Why We Believe in the Unbelievable* (London: HarperOne, 2009), p. 12.

31. Tinbergen, N., "On War and Peace in Animals and Man: An Ethologist's Approach to the Biology of Aggression," *Science* 160 (1968): 1411–18, p. 1416.

32. On preconscious processes: Wilson, T. D., *Strangers to Ourselves: Discovering the Adaptive Unconscious* (Cambridge, MA: Belknap Press, 2004); Kahneman, D., *Thinking, Fast and Slow* (London:

Allen Lane, 2011); Frank, R. H., *Passions Within Reason: The Strategic Role of the Emotions* (New York: Norton, 1988). On emotional input to rational decision-making see Damasio, *Descartes Error: Emotion, Reason and the Human Brain*; McDermott, R., "The Feeling of Rationality: The Meaning of Neuroscientific Advances for Political Science," *Perspectives on Politics* 2, no. 4 (2004): 691–706.

33. Caria, A. et al., "Regulation of Anterior Insular Cortex Activity Using Real-Time fMRI," *NeuroImage* 35, no. 3 (2007): 1238–46.

34. Livnat, A. and Pippenger, N., "An Optimal Brain Can Be Composed of Conflicting Agents," 103, no. 9 (2006): 3198–202.

35. Kapogiannis, D. et al., "Cognitive and Neural Foundations of Religious Belief," *Proceedings of the National Academy of Sciences (PNAS)* 106 (2009): 4876–81. See also Schjoedt, U. et al., "Highly Religious Participants Recruit Areas of Social Cognition in Personal Prayer," *Social Cognitive and Affective Neuroscience* 4, no. 2 (2009): 199–207.

36. Thayer, B. A., "Thinking About Nuclear Deterrence: Why Evolutionary Psychology Undermines its Rational Actor Assumptions," *Comparative Strategy* 26, no. 4 (2007): 311–23, p. 315.

37. McCullough and Carter, "Waiting, Tolerating, and Cooperating: Did Religion Evolve to Prop Up Humans' Self-Control Abilities," p. 424. McCullough and Carter were particularly interested in the *interaction* between the psychology of self-regulation and supernatural agent beliefs, and whether people's mental capacities for "waiting, tolerating, and cooperating" may be better exploited by some types of religious beliefs rather than others (a "gene-culture coevolution" scenario in which our underlying cognitive mechanisms have predisposed us to certain cultural beliefs, or alternatively certain cultural beliefs worked better or worse because of the underlying cognitive structures on which they stand). The coevolutionary process may then have refined religious cognition, via cultural selection, precisely because of its power in promoting self-control.

38. Ibid., p. 428. Italics in original.

39. Carter, E. C., McCullough, M. E., and Carver, C. S., "The Mediating Role of Monitoring in the Association of Religion With

Self-Control," *Social Psychological and Personality Science* 3, no. 6 (2012): 691–97.

40. Csikszentmihalyi, Mihaly, "How to be Good," *Washington Post*, October 15, 2006, p. BW13.

41. Johnson and Bering, "Hand of God, Mind of Man"; Johnson, "Why God Is the Best Punisher."

42. Shakespeare, William, *Measure for Measure*, Act II, Scene I.

43. On moralistic behavior in humans: Alexander, R. D., *The Biology of Moral Systems* (Aldine, NY: Hawthorne, 1987); Haidt, J., "The New Synthesis in Moral Psychology," *Science* 316, no. 5827 (2007): 998–1002; Hauser, M. D., *Moral Minds: How Nature Designed our Universal Sense of Right and Wrong*, 1st ed. (New York: Ecco, 2006); Schloss, J. P., "Darwinian Explanations of Morality: Accounting for the Normal but not the Normative," in *Understanding Moral Sentiments: Darwinian Perspectives?*, ed. Putnam, H., Neiman, S., and Schloss, J. P. (New Brunswick, NJ: Transaction Publishers, 2014); Trivers, R. L., "The Evolution of Reciprocal Altruism," *Quarterly Review of Biology* 46, no. 1 (1971): 35–57.

44. Schloss, J. P., and Murray, M., "Evolutionary Accounts of Belief in Supernatural Punishment: A Critical Review," *Religion, Brain & Behavior (RBB)* 1, no. 1 (2011): 46–99, p. 61.

45. Haselton, M. G., and Nettle, D., "The Paranoid Optimist: An Integrative Evolutionary Model of Cognitive Biases," *Personality and Social Psychology Review* 10, no. 1 (2006): 47–66; Johnson, D. D. P., and Fowler, J. H., "The Evolution of Overconfidence," *Nature* 477, no. 7364 (2011): 317–20; McKay, R. T., and Dennett, D. C., "The Evolution of Misbelief," *Behavioral and Brain Sciences* 32, no. 6 (2009): 493–561; Nesse, R. M., "Natural Selection and the Regulation of Defenses: A Signal Detection Analysis of the Smoke Detector Problem," *Evolution and Human Behavior* 26, no. 1 (2005): 88–105; Nesse, R. M., "The Smoke Detector Principle: Natural Selection and the Regulation of Defensive Responses," *Annals of the New York Academy of Sciences* 935 (2001): 75–85; Johnson, D. D. P. et al., "The Evolution of Error: Error Management, Cognitive Constraints, and Adaptive Decision-Making Biases," *Trends in Ecology & Evolution* 28, no. 8 (2013): 474–81.

46. Johnson, "The Error of God."

47. Johnson, "Why God Is the Best Punisher."
48. On cognitive defaults: Boyer, P., *Religion Explained: The Evolutionary Origins of Religious Thought* (New York: Basic Books, 2001); Atran, S., and Norenzayan, A., "Religion's Evolutionary Landscape: Counterintuition, Commitment, Compassion, Communion," *Behavioural and Brain Sciences* 27, no. 6 (2004): 713–30. The implicit association study is Purzycki, B. et al., "What Does God Know? Supernatural Agents' Perceived Access to Socially Strategic and Nonstrategic Information," *Cognitive Science* 36, no. 5 (2012): 846–69.
49. Superstitious beliefs helping problem solving: Damisch, L., Stoberock, B., and Mussweiler, T., "Keep Your Fingers Crossed! How Superstition Improves Performance," *Psychological Science* 21, no. 7 (2010): 1014–20. Basketball study: Buhrmann, H. G., and Zaugg, M. K., "Superstitions Among Basketball Players: An Investigation of Various Forms of Superstitious Beliefs and Behavior Among Competitive Basketballers at the Junior High School to University Level," *Journal of Sport Behavior* 4, no. 4 (1981): 163–74. Foster, K. R., and Kokko, H., "The Evolution of Superstitious and Superstitious-Like Behaviour," *Proceedings of the Royal Society B: Biological Sciences* 276, no. 1654 (2009): 31–37. David Sloan Wilson quote: Reilly, Michael, "God's Place in a Rational World," *New Scientist*, November 10, 2007.
50. Dawkins, R., *The God Delusion* (New York: Houghton Mifflin, 2006), p. 222.

CHAPTER 7

1. Locke, J., *A Letter Concerning Toleration*, eds. Horton, J., and Mendus, S. (Abingdon, UK: Routledge, 1991/1689), p. 47.
2. Philbrick, N., *Mayflower: A Story of Courage, Community, and War* (New York: Viking, 2006), pp. 5, 7.
3. Ibid., pp. 87–88.
4. Konner, M., *The Evolution of Childhood: Relationships, Emotion, Mind* (Cambridge, MA: Belknap Press, 2011), p. 670; Philbrick, *Mayflower*, p. 29 ("a Puritan believed").

5. Norenzayan, A., *Big Gods: How Religion Transformed Cooperation and Conflict* (Princeton, NJ: Princeton University Press, 2013), p. 45.

6. Hare, J. E., "Moral Motivation," in *Games, Groups, and the Global Good*, ed. Levin, S. A. (Berlin: Springer, 2009); Locke, J., *A Letter Concerning Toleration*, eds. Horton, J., and Mendus, S. (Abingdon, UK: Routledge, 1991/1689), p. 47.

7. Durkheim, É., *The Elementary Forms of Religious Life*, trans. Cosman, C. (Oxford: Oxford University Press, 2001/1912), p. 46; Wilson, D. S., *Darwin's Cathedral: Evolution, Religion, and the Nature of Society* (Chicago: University of Chicago Press, 2002).

8. Norenzayan, *Big Gods*; Whitehouse, H., *Modes of Religiosity: A Cognitive Theory of Religious Transmission* (Walnut Creek, CA: Altamira Press, 2004); Wilson, *Darwin's Cathedral*.

9. Dawkins, R., *The God Delusion* (New York: Houghton Mifflin, 2006), p. 168.

10. Hartung, J., "Love Thy Neighbor: The Evolution of In-Group Morality," *Skeptic* 3, no. 4 (1995): 86–99; Wright, R., *The Evolution of God* (New York: Little, Brown, 2009).

11. Hadnes, M., and Schumacher, H., "The Gods Are Watching: An Experimental Study of Religion and Traditional Belief in Burkina Faso," *Journal for the Scientific Study of Religion* 51, no. 4 (2012): 689–704, pp. 689–90; Mbiti, *African Religions and Philosophy*.

12. Hadnes and Schumacher, "The Gods Are Watching," p. 689.

13. Ostrom, E., *Governing the Commons: The Evolution of Institutions for Collective Action* (Cambridge: Cambridge University Press, 1990).

14. Wilson, David Sloan, "Is Religion Useful? A Test Involving Common Pool Resource Groups," *Evolution: This View of Life*, March 18, 2013. www.thisviewoflife.com.

15. Johnson, D. D. P., "God's Punishment and Public Goods: A Test of the Supernatural Punishment Hypothesis in 186 World Cultures," *Human Nature* 16, no. 4 (2005): 410–46.

16. Warrior, V. M., *Roman Religion* (Cambridge: Cambridge University Press, 2006), p. 40; Van De Mieroop, M., *King Hammurabi of Babylon: A Biography* (Oxford: Blackwell Publishing, 2004).

17. Sosis, R., and Ruffle, B., "Religious Ritual and Cooperation: Testing for a Relationship on Israeli Religious and Secular Kibbutzim," *Current Anthropology* 44, no. 5 (2003): 713–22; Sosis, R., and

Bressler, E. R., "Cooperation and Commune Longevity: A Test of the Costly Signaling Theory of Religion," *Cross-Cultural Research* 37, no. 2 (2003): 211–39; Sosis, R., and Ruffle, B., "Ideology, Religion, and the Evolution of Cooperation: Field Tests on Israeli Kibbutzim," *Research in Economic Anthropology* 23 (2004): 89–117; Fishman, A., and Goldschmidt, Y., "The Orthodox Kibbutzim and Economic Success," *Journal for the Scientific Study of Religion* 29, no. 4 (1990): 505–11.

18. Barro, R., and McCleary, R., "Religion and Economic Growth Across Countries," *American Sociological Review* 68, no. 5 (2003): 760–81; Atkinson, Q. D., and Bourrat, P., "Beliefs about God, the Afterlife and Morality Support the Role of Supernatural Policing in Human Cooperation," *Evolution and Human Behavior* 32, no. 1 (2011): 41–49; Lanman, J., "No Longer 'Walking the Walk': The Importance of Religious Displays for Transmission and Secularization," *Journal of Contemporary Religion* 27, no. 1 (2012): 49–65.

19. Shariff, A. F., and Rhemtulla, M., "Divergent Effects of Beliefs in Heaven and Hell on National Crime Rates," *Public Library of Science ONE (PLoS ONE)* 7, no. 6 (2012): e39048.

20. Zuckerman, P., *Society without God: What the Least Religious Nations Can Tell Us About Contentment* (New York: NYU Press, 2008).

21. Earhart, H. B., ed. *Religious Traditions of the World* (New York: Harper Collins, 1993); Rossano, M. J., "Supernaturalizing Social Life: Religion and the Evolution of Human Cooperation," *Human Nature* 18, no. 3 (2007): 272–94, p. 282.

22. Raphael, J. C., *Mutawas: Saudi Arabia's Dreaded Religious Police* (Mumbai, India: Turtle Books, 2011). The estimate of numbers comes from "Saudi Arabia: Vicious about Virtue," *The Economist*, June 23, 2007.

23. *The Economist*, "Saudi Arabia."

24. Harrison, E., Bromiley, G., and Henry, C., eds., *Wycliffe Dictionary of Theology* (Peabody: Hendrickson, 1960), p. 430.

25. Harris, S., *The End of Faith: Religion, Terror, and the Future of Reason*, 1st ed. (New York: Norton, 2004), pp. 81–87.

26. Dyson, Freeman, "The Prisoner's Dilemma," *Institute Letter*, Princeton Institute of Advanced Study, Fall 2012.
27. Topalli, V., Brezina, T., and Bernhardt, M., "With God on My Side: The Paradoxical Relationship Between Religious Belief and Criminality Among Hardcore Street Offenders," *Theoretical Criminology* 17, no. 1 (2013): 49–69.
28. Dawkins, *The God Delusion*; Zuckerman, *Society without God*.
29. Hirschi, T., and Stark, R., "Hellfire and Delinquency," *Social Problems* 17, no. 2 (1969): 202–13.
30. Baier, C. J., and Wright, B. R. E., ""If You Love Me, Keep My Commandments": A Meta-Analysis of the Effect of Religion on Crime," *Journal of Research in Crime and Delinquency* 38, no. 1 (2001): 3–21, p. 3.
31. Wright, *The Evolution of God*.
32. Schloss, J. P., and Murray, M., "Evolutionary Accounts of Belief in Supernatural Punishment: A Critical Review," *Religion, Brain & Behavior (RBB)* 1, no. 1 (2011): 46–99.
33. Hultkrantz, A., *The Religions of the American Indians* (Berkeley: University of California Press, 1967), p. 187. On fairness in major world religions: Baumard, N., and Boyer, P., "Explaining Moral Religions," *Trends in Cognitive Sciences* 17, no. 6 (2013): 272–80. Harrison, Bromiley, and Henry, *Wycliffe Dictionary of Theology*, p. 430 ("The *lex talionis* of").
34. Dudley, M. K., *A Hawaiian Nation I: Man, Gods, and Nature* (Kapolei, HI: Na Kane O Ka Malo Press, 2003), pp. 71, 110.
35. Boehm, C., *Hierarchy in the Forest: The Evolution of Egalitarian Behavior* (Cambridge, MA: Harvard University Press, 2001); Lee, R. B., and Daly, R., eds., *The Cambridge Encyclopedia of Hunters and Gatherers* (Cambridge: Cambridge University Press, 2004).
36. Bulbulia, J., "Meme Infection or Religious Niche Construction? An Adaptationist Alternative to the Cultural Maladaptationist Hypothesis," *Method and Theory in the Study of Religion* 20, no. 1 (2008): 67–107, p. 102.
37. Weber, M., *The Sociology of Religion* (Berkeley: University of California Press, 1922/1978). See Section 7, "Caste Taboo,

Vocational Caste Ethics, and Capitalism" in the chapter "Magic and Religion."

38. Henrich, J., "The Evolution of Costly Displays, Cooperation and Religion: Credibility Enhancing Displays and Their Implications for Cultural Evolution," *Evolution and Human Behavior* 30 (2009): 244–60. For some discussion of the implications for religion, see Holmes, Bob, "Suffering For Your Beliefs Makes Others Believe Too," *New Scientist*, May 30, 2009.

39. David Sloan Wilson quote in Bartlett, Tom, "Dusting Off God," *The Chronicle of Higher Education*, August 13, 2013.

40. Wright, *The Evolution of God*; Ensminger, J., "Transaction Costs and Islam: Explaining Conversion in Africa," *Journal of Institutional and Theoretical Economics* 152, no. 1 (1997): 4–29.

41. Hartberg et al., Y., "Supernatural Monitoring and Sanction in Community Based Resource Management," *Religion, Brain & Behavior* (in press; published online ahead of print version by Taylor & Francis, October 13, 2014, DOI:10.1080/2153599X.2014.959547).

42. Weber, M., *Economy and Society* (Berkeley: University of California Press, 1978). See Section 4, "Pantheon and Functional Gods" in the chapter "The Origins of Religion." Norenzayan, *Big Gods*. See also Peregrine, P. N., "The Birth of the Gods Revisited: A Partial Replication of Guy Swanson's (1960) Cross-Cultural Study of Religion," *Cross-Cultural Research* 30, no. 1 (1995): 84–112; Swanson, G. E., *The Birth of the Gods* (Ann Arbor: University of Michigan Press, 1960).

43. Baron-Cohen, S., "The Evolution of a Theory of Mind," in *The Descent of Mind: Psychological Perspectives on Hominid Evolution*, ed. Corballis, M. C., and Lea, S. E. G. (Oxford: Oxford University Press, 1999); Bouzouggar, A. et al., "82,000-Year-Old Shell Beads from North Africa and Implications for the Origins of Modern Human Behavior," *Proceedings of the National Academy of Sciences* 104, no. 24 (2007): 9964–69; Corballis, M. C., *The Recursive Mind: The Origins of Human Language, Thought, and Civilization* (Princeton, NJ: Princeton University Press, 2011); Ronen, A., "The Oldest Burials and Their Significance," in *African Genesis: Perspectives on Hominin Evolution*, ed. Reynolds, S. C., and Gallagher, A. (New York: Cambridge University Press, 2012).

44. On the impact of agriculture, see Diamond, J., *Guns, Germs and Steel: A Short History of Everybody for the Last 13,000 Years* (London: Vintage, 1998). On the possibility that religion advanced society, see Wason, P., "Religion, Status and Leadership in Neolithic Avebury: An Example of the Cauvin-Stark 'Religion Drives Innovation' Hypothesis?," *Evolution of Religion, Hawaii Conference 2007* (2007); Whitehouse, H., and Hodder, I., "Modes of Religiosity at Çatalhöyük," in *Religion in the Emergence of Civilization: Çatalhöyük as a Case Study*, ed. Hodder, I. (New York: Cambridge University Press, 2010).

45. For the original distinction made between punishment avoidance and cooperation enhancement, see Schloss and Murray, "Evolutionary Accounts of Belief in Supernatural Punishment." For a recent exposition of cooperation enhancement see Norenzayan, *Big Gods*. Note that punishment avoidance is a misnomer, because as we saw in Chapter 5, what people are trying to avoid is a range of possible forms of retaliation and reputational damage, not just punishment.

46. Johnson, D. D. P., "Why God is the Best Punisher," *Religion, Brain & Behavior (RBB)* 1, no. 1 (2011): 77–84.

47. On problems with the cooperation enhancement approach, see Johnson, "Why God is the Best Punisher"; Johnson, D. D. P., "Big Gods, Small Wonder: Supernatural Punishment Strikes Back," *Religion, Brain & Behavior* (in press; published online ahead of print version by Taylor & Francis, July 22, 2014, DOI: 10.1080/2153599X.2014.928356). On the role of war: Alexander, R. D., *The Biology of Moral Systems* (Aldine, NY: Hawthorne, 1987); Bowles, S., "Group Competition, Reproductive Leveling, and the Evolution of Human Altruism," *Science* 314 (2006): 1569–72; Gat, A., *War in Human Civilization* (Oxford: Oxford University Press, 2006); Keeley, L. H., *War Before Civilization: The Myth of the Peaceful Savage* (Oxford: Oxford University Press, 1996); LeBlanc, S., and Register, K. E., *Constant Battles: The Myth of the Peaceful, Noble Savage* (New York: St. Martin's Press, 2003); Roes, F. L., and Raymond, M., "Belief in Moralizing Gods," *Evolution and Human Behavior* 24, no. 2 (2003): 126–35.

48. Richerson, P. J., and Boyd, R., *Not By Genes Alone: How Culture Transformed Human Evolution* (Chicago: University Of Chicago

Press, 2004); Mesoudi, A., *Cultural Evolution: How Darwinian Theory Can Explain Human Culture and Synthesize the Social Sciences* (Chicago: University of Chicago Press, 2011).

49. Dawkins, *The God Delusion*, p. 201.

50. Maynard Smith, J., and Szathmary, E., *The Major Transitions of Life* (New York: W.H. Freeman, 1995).

51. Cosmides, L., and Tooby, J., "Cognitive Adaptations for Social Exchange," in *The Adapted Mind: Evolutionary Psychology and the Generation of Culture*, ed. Barkow, J. H., Cosmides, L., and Tooby, J. (New York: Oxford University Press, 1992).

52. Olson, M., *The Logic of Collective Action: Public Goods and the Theory of Groups* (Cambridge, MA: Harvard University Press, 1965).

53. Madison, James, "The Structure of the Government Must Furnish the Proper Checks and Balances Between the Different Departments From the New York Packet," *Federalist* no. 51 (1788), Friday, February 8, 1788.

CHAPTER 8

1. "Woman Auctions Father's Ghost on eBay," *Associated Press*, December 4, 2004 (accessed at CNN.com); "Ghost Cane Nets $65,000 on eBay," *Associated Press*, (accessed at NBCnews.com). The winning bidder was GoldenPalace.com, an online casino based in Antigua, which had also bought a grilled-cheese sandwich deemed to bear the image of the Virgin Mary for $28,000. Due to eBay rules, which require that you have to actually sell something physical (otherwise delivery cannot be confirmed), Mary Anderson threw her father's cane into the deal.

2. Lawrence, B. B., *The Complete Idiot's Guide to Religions Online* (New York: Macmillan, 2000).

3. Bremmer, J. N., "Atheism in Antiquity," in *The Cambridge Companion to Atheism*, ed. Martin, M. (Cambridge: Cambridge University Press, 2007), p. 2.

4. For these and more examples see Zuckerman, P., *Society Without God: What the Least Religious Nations Can Tell Us About Contentment* (New York: NYU Press, 2008).

5. Dawkins, *The God Delusion*; Zuckerman, *Society Without God*. Hauser, M., and Singer, P., "Morality Without Religion," *Free Inquiry* 26, no. 1 (2006): 18–19; Dawkins, *The God Delusion*, see pp. 226–27.

6. Pew Forum, 2008, "Being Good for Goodness' Sake?" www.pewforum.org/Being-Good-for-Goodness-Sake.aspx.

7. Gervais, W. M., Shariff, A. F., and Norenzayan, A., "Do You Believe in Atheists? Distrust Is Central to Anti-Atheist prejudice," *Journal of Personality and Social Psychology* 101, no. 6 (2011): 1189–206, p. 1189.

8. Ibid, p. 1189.

9. Nietzsche, F., *On the Geneology of Morals and Ecce Homo*, trans. Kaufman, W. (New York: Vintage Books, 1967), pp. 90–91; Barro, R., and McCleary, R., "Religion and Economic Growth Across Countries," *American Sociological Review* 68, no. 5 (2003): 760–81; Shariff, A. F., and Rhemtulla, M., "Divergent Effects of Beliefs in Heaven and Hell on National Crime Rates," *Public Library of Science ONE (PLoS ONE)* 7, no. 6 (2012): e39048.

10. Barrett, David, "One Surveillance Camera for Every 11 People in Britain, Says CCTV Survey," *The Telegraph*, July 10, 2013.

11. Guillermoprieto, Alma, "Troubled Spirits," *National Geographic*, May 2010. See also Chesnut, R. A., *Devoted to Death: Santa Muerte, the Skeleton Saint* (New York: Oxford University Press, 2012).

12. Guillermoprieto, "Troubled Spirits."

13. Guillermoprieto, "Troubled Spirits."

14. Chestnut, *Devoted to Death*, p. 7; Guillermoprieto, "Troubled Spirits."

15. Yang, F., *Religion in China: Survival and Revival under Communist Rule* (Oxford: Oxford University Press, 2011). Quotes are from "Briefing: Religion in China: When Opium Can be Benign," *The Economist*, February 3, 2007, pp. 23–25.

16. *The Economist*, "Briefing," pp. 23–25, 25.

17. Storm, I., "Halfway to Heaven: Four Types of Fuzzy Fidelity in Europe," *Journal for the Scientific Study of Religion* 48, no. 4 (2009): 702–18.

18. Norris, P., and Inglehart, R., *Sacred and Secular: Religion and Politics Worldwide* (Cambridge: Cambridge University Press, 2004);

Shah, T. S., and Toft, M. D., "Why God is Winning," *Foreign Policy* 155, no. July/August (2006): 39–43.

19. Mead, W. R., "God's Country?," *Foreign Affairs* 85, no. 5 (2006): 24–43; Rosenau, W., "US Counterterrorism Policy," in *How states fight terrorism: policy dynamics in the West*, ed. Zimmermann, D., and Wenger, A. (Boulder, CO: Lynne Rienner Publishers, 2007), p. 134.

20. Almond, G. A., Appleby, R. S., and Sivan, E., *Strong Religion: The Rise of Fundamentalisms Around the World* (Chicago: Chicago University Press, 2003); Kepel, G., *The Revenge of God* (University Park: Pennsylvania State University Press, 1994), p. 2.

21. Lane, J. E., and Ersson, S. O., *Culture and Politics: A Comparative Approach* (Aldershot, UK: Ashgate, 2005); Martin, W., *With God on Our Side: The Rise of the Religious Right in America* (New York: Broadway Books, 1997).

22. Kohut, A. et al., *The Diminishing Divide: Religion's Changing Role in American Politics* (Washington, DC: Brookings Institution Press, 2000), p. 29.

23. Norris and Inglehart, *Sacred and Secular.*

24. Huntington, S. P., *The Clash of Civilizations and the Remaking of World Order* (New York, NY: Touchstone, 1997).

25. Pascal, B., *Pensées and Other Writings*, ed. Levi, A., trans. Levi, H., Oxford World's Classics (Oxford: Oxford University Press, 2008/1669).

26. Quoted in Dawkins, *The God Delusion*, p. 228.

27. Robinson, P. H., and Darley, J. M., "Does Criminal Law Deter? A Behavioural Science Investigation," *Oxford Journal of Legal Studies* 24, no. 2 (2004): 173–205.

28. Bering, J. M., *The God Instinct: The Psychology of Souls, Destiny and the Meaning of Life* (London: Nicolas Brealey, 2010), p. 200.

29. Hood, B. M., *SuperSense: Why We Believe in the Unbelievable* (London: HarperOne, 2009), p. 60; Wilson, *Consilience: The Unity of Knowledge*, p. 292; McCauley, R. N., *Why Religion is Natural and Science is Not* (Oxford: Oxford University Press, 2011).

30. Harris, S., *The End of Faith: Religion, Terror, and the Future of Reason*, 1st ed. (New York: Norton, 2004), p. 149.

31. Editorial, "The Trouble With Reason," *New Scientist*, November 10, 2007.

CHAPTER 9

1. Newton, J., and Cowper, W., 1779, *Olney Hymns*.
2. Goddard, J., "Buried for 27 Days: Haiti Earthquake Survivor's Amazing Story," *The Telegraph* (London), March 28, 2010.
3. Goddard, "Buried for 27 Days"; Tedmanson, Sophie, "Man 'Buried Under Rubble for Four Weeks' Found Alive in Haiti," *The Times* (London), February 9, 2010.
4. Rossano, M. J., "Supernaturalizing Social Life: Religion and the Evolution of Human Cooperation," *Human Nature* 18, no. 3 (2007): 272–94, p. 273.
5. Andre, C., and Velasquez, M., "Just World Theory," *Issues in Ethics* 3, no. 2 (1990), http://www.scu.edu/ethics/publications/iie/v3n2/justworld.html; Hafer, C. L., and Bègue, L., "Experimental Research on Just-World Theory: Problems, Developments, and Future Challenges," *Psychological Bulletin* 131, no. 1 (2005): 128–67; Lerner, M. J., "The Justice Motive: Some Hypotheses as to its Origins and Forms," *Journal of Personality* 45, no. 1 (1977): 1–52.
6. For theories on the role of ritual, see Sosis, R., and Alcorta, C., "Signaling, Solidarity, and the Sacred: The Evolution of Religious Behavior," *Evolutionary Anthropology* 12, no. 6 (2003): 264–74; Irons, W., "Religion as a Hard-to-Fake Sign of Commitment," in *Evolution and the Capacity for Commitment*, ed. Nesse, R. (New York: Russell Sage Foundation., 2001); Bulbulia, J., "Meme Infection or Religious Niche Construction? An Adaptationist Alternative to the Cultural Maladaptationist Hypothesis," *Method and Theory in the Study of Religion* 20, no. 1 (2008): 67–107.
7. Boyer, P., and Liernard, P., "Ritual Behavior in Obsessive and Normal Individuals," *Current Directions in Psychological Science* 17, no. 4 (2008): 291–94, p. 292.
8. Dawkins, R., *The God Delusion* (New York: Houghton Mifflin, 2006), p. 222.

INDEX